MULTIPLE-BEAM INTERFEROMETRY OF SURFACES AND FILMS

BY

S. TOLANSKY
PROFESSOR OF PHYSICS, ROYAL HOLLOWAY COLLEGE
THE UNIVERSITY OF LONDON

DOVER PUBLICATIONS, INC.
NEW YORK

Published in Canada by General Publishing Company, Ltd., 30 Lesmill Road, Don Mills, Toronto, Ontario.

Published in the United Kingdom by Constable and Company, Ltd., 10 Orange Street, London WC 2.

This Dover edition, first published in 1970, is an unabridged and unaltered republication of the work originally published in 1948. It is reprinted by special arrangement with Oxford University Press, Inc., publisher of the original edition.

International Standard Book Number: 0-486-62215-0
Library of Congress Catalog Card Number: 72-125910

Manufactured in the United States of America
Dover Publications, Inc.
180 Varick Street
New York, N. Y. 10014

PREFACE

This book describes in detail a number of optical interference techniques developed by the writer since 1942. These experimental procedures have already led to results of value in a number of different fields. They are partly refinements of established classical techniques which are over fifty years old, but it is just these very refinements that constitute an advance and permit of a precision in observation and in measurement much ahead of the earlier work.

The basis of all the methods described is the correct use of multiple-beam interference. By this means it is possible to study surface topography, and thin films, down to molecular dimensions, and indeed although only light waves are used, sizes of crystal lattice spacings can be measured in favourable instances.

A notable feature is the simplicity of means with which such high magnifications (of the order of half a million) and such high resolution (in some cases only 5 A.U.) are achieved. In a sense the techniques described give results complementary to those of the electron microscope, for whilst the latter is most frequently used to magnify areas, the interference method gives great magnification in height and depth. Although recent elaborate electron-microscope stereoscopic methods claim to measure heights of the order of 100 A.U., yet the complexity (and expense) of such a procedure is not to be compared with the simplicity of the optical method, which in any case is more than 10 times as sensitive, and surpasses too the elaborate shadow-casting technique of electron microscopy.

Although it is but five years since the first of these interference techniques was developed, the methods described have already produced results of value in the study of crystal surface topography, the mechanism of cleavage, optical and dielectric properties of mica, optical properties of metallic films, surface characteristics of plastics, optical properties of thin films, etc.

Further lines of development now being pursued concern

metallic polish and optical properties of thin films. Since the technique clearly has applications in chemistry, crystallography, crystal physics, and metallurgy, this book is definitely not addressed to the optics specialist. For this reason some space has been devoted to the elementary optics of interference and the derivation of Airy's formula is included as an appendix to Chapter II.

I take this welcome opportunity of expressing thanks to my research students for assisting in preparing the many photographs used as illustrations. They have actively contributed to the development of the subject and I have freely made use of work carried out with them jointly under my direction. Especial contributions have been made by P. G. Morris with mica, W. L. Wilcock with diamond, A. Khamsavi with selenite, calcite, and thin films, W. K. Donaldson with light filters and reflecting films, J. Brossel with properties of fringes, and A. Faust with plastics.

S. T.

May 1947

CONTENTS

CHAPTER I

TWO-BEAM INTERFERENCE

Fizeau fringes

It is the object of this treatise to show how precision optical interference methods can be used for the study of (1) the surface topography of reasonably smooth surfaces, including crystals, plastics, metals, etc.; (2) some properties of thin films; (3) certain optical properties of metals.

The methods described are capable of high precision with the simplest of means, and measurements on features of only molecular dimensions can be carried out. These interference techniques employ what will be called *localized multiple-beam fringes*, which are a sensitive optical refinement of the ordinary classical 'two-beam' interference so widely used in optical and engineering workshops for the examination of glass and metal surfaces and for many metrological purposes.

The study of the contour of an approximately plane smooth surface by using optical interference dates back to 1862, in which year Fizeau [1] introduced his celebrated procedure. The use of 'Fizeau' fringes is so widely known that it is necessary to refer to it only briefly.

It is shown in elementary books that when optical interference takes place in a thin transparent wedge of refractive index μ, straight-line fringes occur at wedge thicknesses t given by $n\lambda = 2\mu t \cos\phi$, in which ϕ is the angle of incidence of the light, λ the wave-length, and n the order of interference. It is normally necessary to view such a wedge in reflection, for then the amplitudes of the two interfering beams are almost identical and the fringe 'visibility' is good. For example, with glass some 4 per cent. of the incident light is reflected from each face and these two beams give fringes which have visibility effectively unity, i.e. there is no light half-way between bright fringes. In transmission, fringes of similar intensity and distribution are superposed upon an intense background and are consequently hardly visible.

With a very thin film (e.g. soap bubble or a thin air film between glass plates) reflected fringes are readily seen with an extended reasonably monochromatic source (e.g. sodium flame), but as the thickness increases, it is necessary to restrict ϕ to a single value, i.e. a critically parallel light beam must be used.

A typical Fizeau fringe set-up is shown in Fig. 1. The source A is a green filtered mercury arc and an image of this is projected

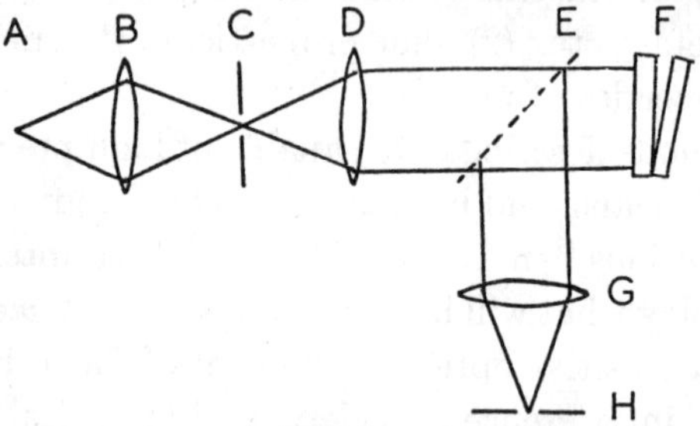

FIG. 1.

by the lens B on to the circular aperture C, of the order of 1 mm. in diameter. C is at the focus of a good lens D. A parallel beam of light passes through the glass plate E and strikes the wedge F at normal incidence. (The wedge angle is so small that incidence is effectively normal on both front and back faces of F.) The light is reflected back from F, forming an auto-collimated image on C when alinement is correct. The glass plate E, which may be half-silvered, reflects light into the good lens G and at the focus of the latter at H is a small aperture behind which the eye is placed and fringes are seen. The plate E leads to considerable loss of light, and an alternative arrangement (which uses one lens only) is that shown in Fig. 2 a, in which a slight deviation from the normal takes place. The source is at A, the eye at B. Fig. 2 b shows a version retaining normal incidence at the expense of light. In this, A and B can be interchanged. The lens focal lengths should be about 50 cm.

Let the wedge be an air film produced by placing the surface PC to be studied close to a standard optical flat QC, then as $\mu = \cos\phi = 1$, fringes will be formed which represent a contour map of the topography of the surface of PC. A fringe represents the locus of points for which n is constant, hence each fringe is the locus of points for which t is constant. The fringes are thus classed as 'fringes of equal thickness' and appear to be localized in the interference film.

On moving along the wedge from one fringe to its neighbour the thickness of the air space has changed by $\lambda/2$. It would at

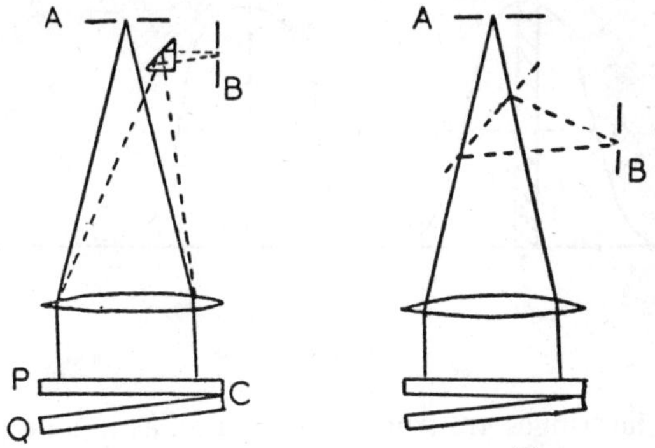

(a) (b)

FIG. 2.

first appear advantageous to use as short a wave-length as possible. This is in fact often not the case, for the reference flat is invariably surfaced in green mercury light. Let it be worked flat, for example to 1/5th of a light-wave at 5,000 A.U. (error 1,000 A.U.). Clearly at 4,000 A.U. this error is $\frac{1}{4}$ of a light wave. The gain in using a shorter wave-length is annulled by the greater relative imperfection of the standard.

Now elementary calculation shows that when two equal beams of light interfere, the intensity within the fringes follows the distribution of a $\cos^2$ curve, in which the light and dark areas are equal. This distribution is shown graphically in Fig. 3. The half-width of the fringe ω is defined as the width AB at half the intensity of the maximum and is clearly $\frac{1}{2}CD$, where CD is the distance between orders. If the fringe maxima can be located to within $\omega/10$, that is within the shaded area, the attainable precision is thus $CD/20 = \lambda/40$. This degree of precision is generally quoted in the literature, but it is clearly an upper limit and also obviously refers to relatively large areas, for small local variations within a fringe width are not likely to be detected at all. Whilst, in the most favourable case, $\lambda/40$ may represent the accuracy in setting on a fringe, the ability to resolve small imperfections is not nearly as good.

Within their limitations, the simple Fizeau fringes afford considerable information about the topographical structure of the

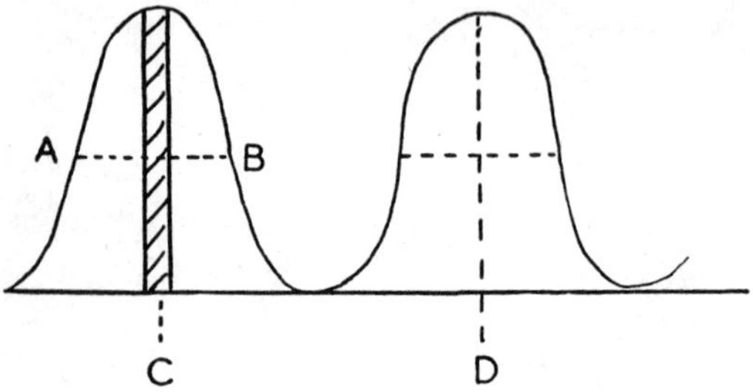

FIG. 3.

surface. The fringes are true contour lines, as in a geographical contour map, the height interval between contours being $\lambda/2$. It is often easy to tell which are hill and which are valley features, for by gently pressing PQ together (Fig. 2a) it is found that the fringes move towards PQ. The region of lowest order is thus determined and higher order means higher t value, so that there is no ambiguity. If the wedge angle cannot be altered, then special methods must be employed. These will be described later.

Newton's rings

A special case of thin film fringes of equal thickness, and one of much historical importance in optics, is that of Newton's rings.† These are obtained by resting the curved face of a long focal length plano-convex lens on a piece of glass which replaces the wedge in Fig. 2b. Since the two surfaces are close together, critical collimation can be dispensed with. An extended source can replace A and the lens can be removed.

It is clear why circular rings arise, for obviously the contours of a spherical lens will be circles. If r is the radius of a given nth

† First discovered by Hooke, but later studied by Newton.

ring and R the radius of curvature of the convex lens surface, it may be shown that

$$\frac{\mu r^2}{R} = n\lambda,$$

if μ is the refractive index of the medium between the glass surfaces.

The radii are proportional to the roots of the natural sequence of numbers 1, 2, 3, 4,.... This holds for successive dark rings, the centre point of contact ($n = 0$) being dark because of phase change at reflection at one of the surfaces. The intensity distribution across the system is a $\cos^2$ distribution and this limits the sensitivity, in the manner already considered. Often the optical components are not in perfect contact, in which case a slightly modified formula must be used.

Haidinger's fringes

A special case of interference arises when t is constant, i.e. the surfaces are plane and parallel. In this case the eye at B in Fig. 2 *a* sees uniform illumination, the intensity of which depends upon μt and is zero when $\mu t = n\lambda$ (zero, because of the phase change of π which takes place on reflection at PC). If, however, the aperture A is removed and an extended source substituted, then interference rings can be seen with an eyepiece at B. In the formula $n\lambda = 2\mu t \cos\phi$, t is now constant and ϕ variable, and clearly for certain values of ϕ, n will be integral, leading to interference. All light falling on the plate at angle ϕ will contribute to this particular fringe. Thus all the light incident on the surface of a cone of semi-angle ϕ will form a fringe of given order n. Clearly interference rings form, and it is easy to show that the ring diameters are proportional to the square roots of the natural sequence of numbers.

As the two interfering beams are parallel (because the two surfaces are parallel), the fringes appear to be at infinity and can be seen by a telescope set on infinity, achieved by introducing an eyepiece at B. Alternatively the rings can be seen with the eye if relaxed on infinity.

These rings, discovered by Haidinger [2] in 1849 using thin slips of mica, have also a $\cos^2$ distribution. Since each fringe corresponds to a given value of ϕ, they are called 'fringes of equal inclination', to distinguish them from the Fizeau 'fringes of equal thickness'.

Twyman-Green interferometer

For many years the Fizeau fringe technique remained the most widely used method for testing optical surfaces. Another important method was developed in 1916 by Twyman and Green

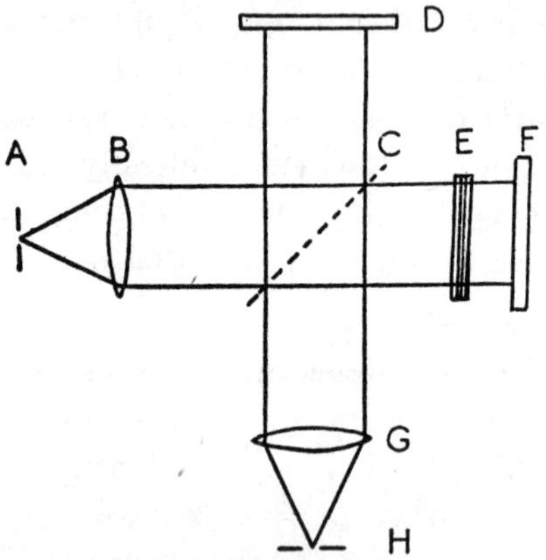

FIG. 4.

[3]. This is similar to the Fizeau method, but employs a Michelson interferometer to divide the beam in the manner shown in Fig. 4. The essential features only are shown. The lens B produces a parallel beam from the point source A. The light falls on the dividing mirror C, beams of equal amplitude being sent normally to the mirrors D and F. The light returns and is recombined by the lens G. At an aperture H are seen fringes of equal thickness. The object to be studied is E, and typical fringe contours appear at H if E is not plane parallel. The arrangement has been adapted to the examination of prisms and lenses. As in the Fizeau arrangement, *two beams* only are used and the fringe distribution is consequently of the $\cos^2$ type, with its restricted applicability.

Multiple-beam fringes

In the next section it will be shown that interference produced by a large number of beams (*multiple-beam interference*) leads to a great increase in precision. It is readily seen by analogy with a grating why this should be so. The spectrum lines produced by a grating consisting of two apertures only are broad, and effectively $\cos^2$ in shape. But with a grating consisting of no more than 50 apertures the secondary maxima between principal orders are very weak and a great sharpening of the main maxima takes place. The principle of multiple-beam interference is the basis of the celebrated Fabry-Perot [4] interferometer (1897), and multiple-beam Fizeau fringes were also used by Benoit, Fabry, and Perot [5] in their classical determination of the metre, but, as will be shown later, the optimum conditions for the best sensitivity were not then achieved.

It seems that Adam Hilger Ltd. were among the first to employ multiple-beam Fizeau fringes for the examination of optical flats. This technique was in use prior to 1929 [6], but again the best arrangement was not realized in practice.

The multiple-beam effect is produced by coating the surfaces producing interference with a high reflecting thin film of silver through which a fraction of the light can be transmitted. The details of technique will be given later.

CHAPTER II

MULTIPLE-BEAM INTERFERENCE

Airy's formula

THE theory of the multiple-beam interference taking place in a plane parallel plate (a flat plate of uniform thickness) was first given by Airy [7] and will be briefly indicated here. Let A, B in Fig. 5 be parallel reflecting surfaces enclosing a medium of refractive index μ. Let fractions of a beam, which has unit ampli-

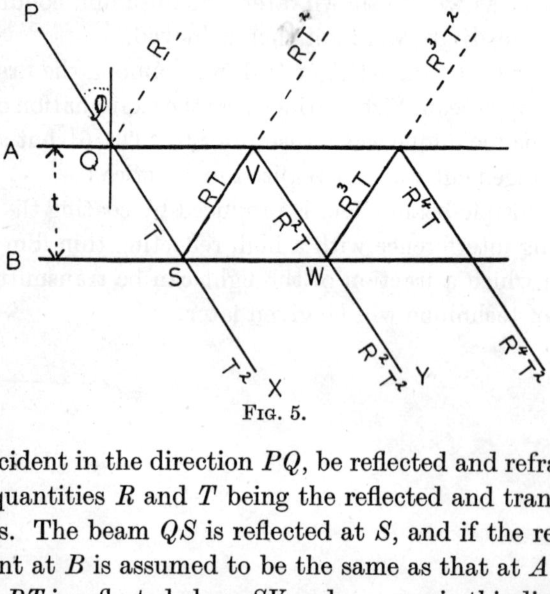

FIG. 5.

tude, incident in the direction PQ, be reflected and refracted at Q, the quantities R and T being the reflected and transmitted fractions. The beam QS is reflected at S, and if the reflection coefficient at B is assumed to be the same as that at A, then a fraction RT is reflected along SV and emerges in this direction.† Thus there are transmitted along SX, WY, etc., parallel rays of intensity

$$T^2, R^2T^2, R^4T^2, \ldots.$$

If these are allowed to fall on a collecting lens, all the beams will combine at the focus of this lens. Now each successive beam lags behind the first beam a distance $2\mu t \cos\phi$, in which ϕ is the angle of incidence and t the distance between the parallel

† To simplify the calculation, refraction effects are disregarded, i.e. AB are to be treated as free unsupported reflecting surfaces.

surfaces. The phase lag δ between successive beams is constant such that

$$\delta = \frac{2\pi}{\lambda} 2\mu t \cos\phi.$$

The Haidinger rings formed at the focus of the collecting lens are modified by the multiple-beam combination. It can readily be shown that the ring diameters are not affected, but that the fringe-intensity distribution changes and is no longer of the $\cos^2$ type.

If the transmitted series of beams of geometrically decreasing intensity (decreasing by factor R^2) and phase increasing arithmetically by δ is summed to infinity, the resulting intensity at any point in the field corresponding to δ can be shown to be (see appendix to this chapter):

$$I = \frac{T^2}{(1-R)^2} \times \frac{1}{1+\{4R/(1-R)^2\}\sin^2\frac{1}{2}\delta}.$$

The quantity $\sin^2\frac{1}{2}\delta$ can only vary from 0 to 1, at which I has maximum and minimum values respectively. When $\sin^2\frac{1}{2}\delta = 0$,

$$I_{\text{max}} = \frac{T^2}{(1-R)^2}.$$

If there is no absorption at the reflecting surfaces, then

$$T = 1-R \quad \text{and} \quad I_{\text{max}} = 1,$$

i.e. the intensity of the fringe maximum is equal to that of the incident light, no matter what the values of R and T.

If, however, a fraction A be absorbed at each surface, then

$$I = A+T+R,$$

and

$$I_{\text{max}} = \left(\frac{T}{T+A}\right)^2 = \left(\frac{1}{1+A/T}\right)^2.$$

This is an important quantity and will be brought back into the discussion later.

When $\sin^2\frac{1}{2}\delta = 1$, and there is no absorption, I is a minimum and has the value

$$I_{\text{min}} = \left[\frac{T^2}{(1-R)^2} \times \frac{1}{1+4R/(1-R)^2}\right] = \frac{T^2}{(1+R)^2} = \left(\frac{1-R}{1+R}\right)^2.$$

With absorption,

$$I_{\min} = \left(\frac{1-R}{1+R} - \frac{A}{1+R}\right)^2,$$

but the ratio $I_{\max}/I_{\min}$ remains $\left(\frac{1+R}{1-R}\right)^2$.

The whole fringe *shape* is thus quite independent of absorption.

I can be written as equal to $\frac{I_{\max}}{1+F\sin^2\frac{1}{2}\delta}$, in which F is $\frac{4R}{(1-R)^2}$ and was called by Fabry the 'coefficient of finesse'.

Effect of varying reflecting coefficients

It was first pointed out by Boulouch [8] that a profound change in the appearance of the interference fringes takes place

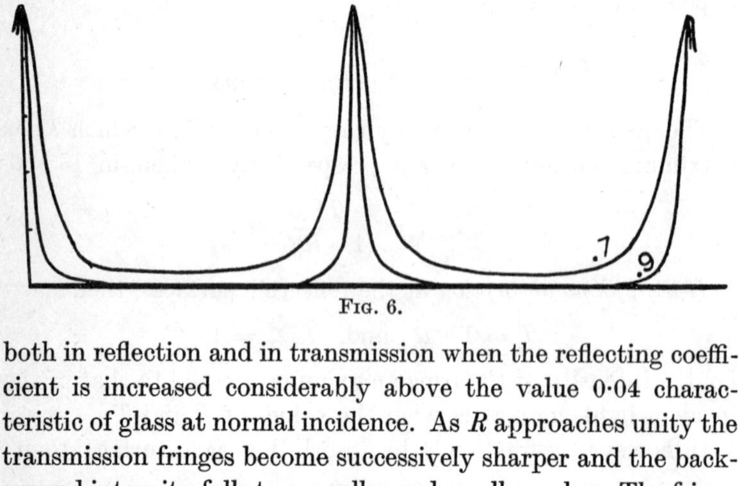

FIG. 6.

both in reflection and in transmission when the reflecting coefficient is increased considerably above the value 0·04 characteristic of glass at normal incidence. As R approaches unity the transmission fringes become successively sharper and the background intensity falls to a smaller and smaller value. The fringe shapes for $R = 0{\cdot}7$ and $0{\cdot}9$ are shown in Fig. 6. Boulouch succeeded in demonstrating the effect, but it remained for Fabry and Perot (loc. cit.) to bring to a full realization the importance of this fact. The increase in reflecting coefficient is obtained by the deposition of a thin film of silver on the surfaces involved. It is necessary to reduce the absorption to a minimum—otherwise the intensity is reduced too far for practical use. Indeed, the technique of silver deposition for interferometry is concerned as much with the production of low absorption as with high reflection. Evaporation techniques have now been developed which permit of reflection coefficients of up to 94 per cent. with about 1 per cent. transmission and 5 per cent. absorption. With such films very sharp fringes result, *provided that all the effective beams are collected with the image-forming lens.*

If the reflecting coefficients of the two surfaces differ and are R_1, R_2, the formulae given above hold if R is replaced everywhere by $\sqrt{(R_1 R_2)}$.

The Fabry-Perot interferometer

The Fabry-Perot interferometer is so well known that a brief survey only will be given here. Fabry and Perot fully developed Boulouch's proposal. Their instrument consists of two high-grade optical flats coated with high reflecting coefficient silver films. The two flats are adjusted parallel and separated (according to requirements) by a distance which usually varies from 1 to 200 mm. in practice. When monochromatic light, from an extended source, is directed on to the interferometer, modified transmission Haidinger fringes are formed. For any angle of incidence the successive multiple reflected emergent beams are all parallel to this incident direction (since the two flats are parallel). If these beams are collected with a lens the Airy summation at the lens focus is automatically secured. Hence in the focal plane appear highly sharpened rings, as shown in Fig. 7 (green mercury line). The ring diameters are proportional to the focal length of the image-forming lens and inversely as $\sqrt{t}$, where t is the separation of the two flats. As is well known, these fringes have considerable practical importance in such fields as metrology, high-resolution spectroscopy, wave-length determination, etc.

The Fabry-Perot fringes appear to be at infinity and have not yet found application in surface studies. However, closely related non-localized fringes have been developed for this purpose and will be described in a later chapter.

The fringes occur at angular diameters given by integral values of n in the expression $n\lambda = 2\mu t \cos\phi$. In effect the sharpening of the rings makes the instrument behave as an *angular filter*.

In the Fabry-Perot interferometer, fringe width, resolving power, and intensity of the background between the maxima are all of importance. In these studies background intensity is of little interest, the important feature being fringe width.

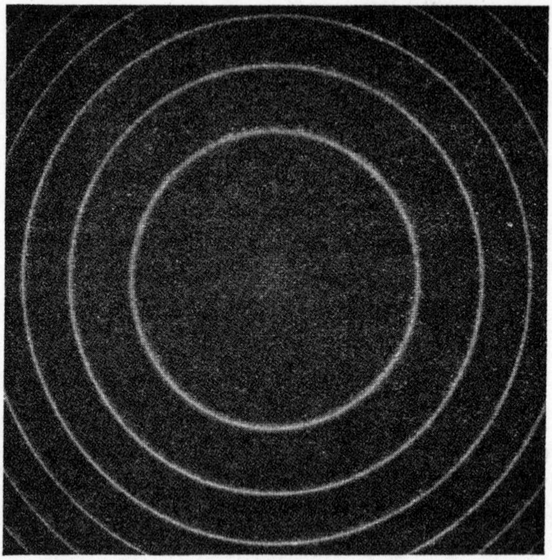

FIG. 7.

Fringe width

It is usual to refer to the *half-width* as the width at half the peak intensity. This is readily calculated from the Airy formula by the simplified approximation given by Burger and van Cittert [9].

Writing
$$I = \frac{I_{\max}}{1+F\sin^2\frac{1}{2}\delta},$$

the fringes are so sharp that over the half-width δ is small enough to replace its sine, hence

$$I = \frac{I_{\max}}{1+\frac{1}{4}F\delta^2} = \frac{I_{\max}}{1+\{R/(1-R)^2\}\delta^2}.$$

At the half-width $I = \frac{1}{2}I_{\max}$ by definition, giving

$$\delta = \frac{1-R}{\sqrt{R}}.$$

Since the fringe falls off symmetrically on either side of the maximum to $\frac{1}{2}I_{max}$, the fringe phase angle corresponding to the fringe half-width is twice the above, i.e. $2\{(1-R)/\sqrt{R}\}$. The phase interval between successive orders is 2π, hence the fringe half-width *as a fraction of an order* is

$$W = \frac{1-R}{\pi\sqrt{R}}.$$

Table I gives approximate values of W for a number of values of R.

TABLE I

R	0·04	0·7	0·8	0·85	0·9	0·925	0·94
W	$\frac{1}{3}$	$\frac{1}{9}$	$\frac{1}{14}$	$\frac{1}{19}$	$\frac{1}{30}$	$\frac{1}{40}$	$\frac{1}{50}$

It is possible to work with $R = 0{\cdot}94$, which gives fringes 25 times narrower than two-beam fringes for which $W = \frac{1}{2}$. The quantity $W = \frac{1}{50}$ enables settings to be made to $\lambda/1000$ if again a tenth of a fringe width is taken to be the setting-error. With the green mercury line ($\lambda = 5460$) this corresponds to the remarkably small value of only $5\frac{1}{2}$ A.U. In fact, owing to the steep fall of intensity, settings can be made to 1/20th of the fringe width, corresponding to an error of less than 3 A.U., when multiple-beam Fizeau fringes are used. This quantity is of atomic dimensions, and therein lies the importance of the technique.

The above calculations apply strictly to plane parallel surfaces in which all the effective beams are used. In practice this cannot be secured because of (*a*) surface imperfections, (*b*) lack of perfect parallelism, (*c*) finite aperture. With the Fabry-Perot interferometer, fringe widths approach the theoretical values only in rare instances. It is particularly to be noted that all the surface imperfections are integrated and cause broadening for every fringe.

Multiple-beam Fizeau fringes [10, 11, 12]

Airy's formula can be written as:

$$I = \frac{I_{max}}{I+F\sin^2\pi\{(2\mu t\cos\phi)/\lambda\}}.$$

If μ is taken to be unity (air gap), then there are three possible variables, ϕ, t, λ, if the value of R is maintained constant.

With t, λ constant we have fringes of equal inclination, i.e. Fabry-Perot fringes.

Clearly with ϕ, λ constant an identical fringe-intensity distribution results and we have multiple-beam fringes of equal thickness, provided some means can be devised by which all the beams are summed.

It will be shown later that with ϕ, t constant and λ variable, another type of fringe arises. This has particularly valuable applications when both t and λ are made to vary simultaneously.

The summation of all the beams in the Fabry-Perot interferometer is easy to achieve, for all the beams (which are parallel) are brought to the focus of the collecting lens, and in theory an Airy distribution should obtain. In practice this is approximated to. In multiple-beam *Fizeau fringes* this is not the case. At normal incidence Fizeau fringes are localized in the interference film, but the multiple beams can only be considered to combine locally under certain critical illumination conditions which will now be examined.

The phase condition

Consider first the mode of formation of localized fringes with two interfering beams (Fig. 8). Let a parallel beam of light between C and I fall at normal incidence on NO, the extreme rays meeting LM at D and J. Consider another ray (normal incidence) from A, which after successive reflection at B and E meets the ray from C at D. At the point D_1 conjugate with D is formed an image of the localized interference fringe.

In like manner the beams $FGHJ$ and IJ interfere at J and form a conjugate fringe image at J_1.

The parallel rays DP, JR meet to form a point image X_1 at the principal focus of the lens, and since all rays incident between C and I also arrive at X_1 no fringe is formed there; only an image. In a similar way DQ and JS form a point image at X_2. In accordance with the Abbe theory of the microscope, X_1 and

X_2 can be regarded as true point sources, leading to fringe formation in the plane $J_1 D_1$.

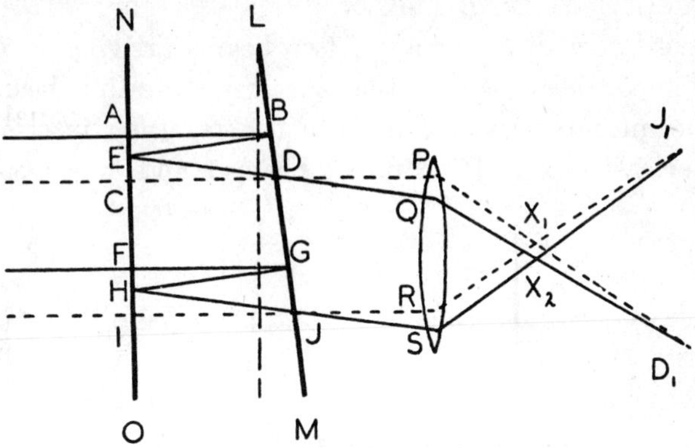

FIG. 8.

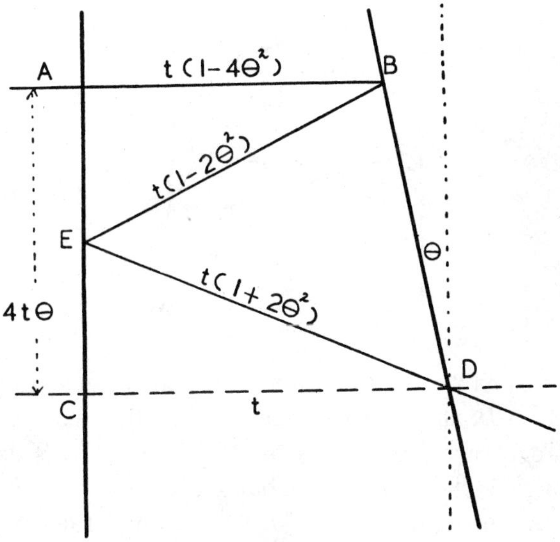

FIG. 9.

Consider now, as shown in Fig. 9, the path difference between the two beams arriving at D. To a rough first approximation, θ being small,

$$ED = t(1+2\theta^2),$$
$$EB = t(1-2\theta^2),$$
$$AB = t(1-4\theta^2),$$

making the path difference $2t-4t\theta^2$. It will be noted that the two beams uniting at D are associated with different points on the same surface, i.e. B and D.

Suppose now that instead of two beams arriving at D, the reflecting coefficient is such that successive multiple beams are of sufficient intensity to influence the resulting interference. Consider, as in Fig. 10, a direct beam 1 and two successive

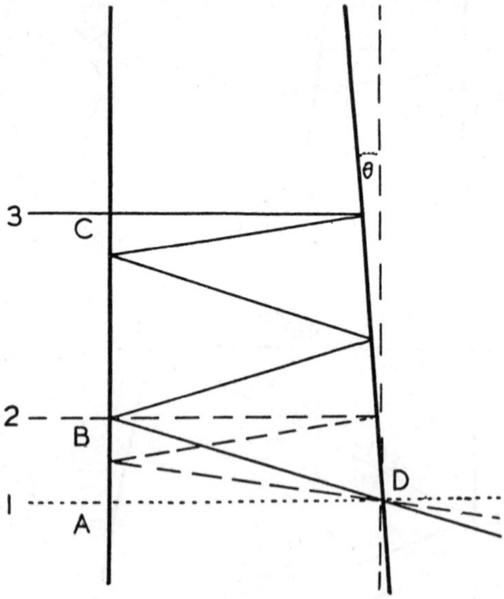

FIG. 10.

multiple beams 2 and 3. Since the incident beams are parallel and meet at D, they must impinge on the first surface at distances progressively farther from A. The beam 2 is deflected first through 2θ then through 4θ, but the beam 3 is deflected first through 2θ, then successively 4θ, 6θ, 8θ, and so on for higher orders.

It is clear that the path difference between successive reflected beams is not constant but alters progressively with the order of reflection. By extending the previous calculation for two beams it is readily shown that to a rough approximation the path difference between the first and nth beams is $2nt-\frac{4}{3}n^3\theta^2 t$. Thus the phases of beams of higher order gradually get out of step with the first beam, and instead of assisting the Airy summation series begin ultimately to oppose it. Indeed, when the retardation is $\lambda/2$, such a beam tends to destroy the condition of sharpness.

A more exact derivation of the phase lag has been carried out by J. Brossell at the Manchester University Laboratory, who gives the following simple general method:

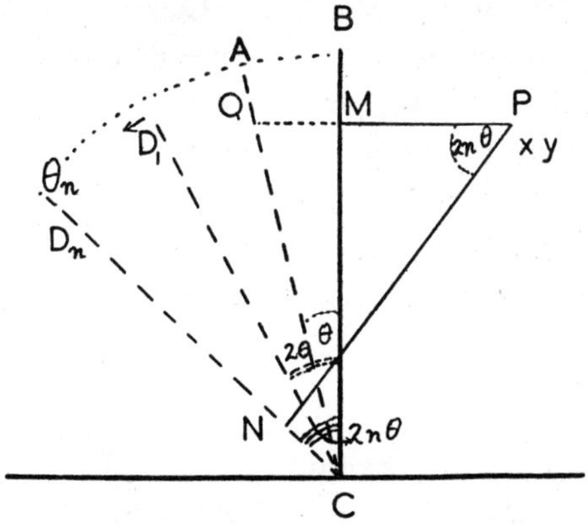

FIG. 11.

Let AC, CB of Fig. 11 represent the wedge and also the wave fronts reflected at each surface. Then $CD,\ldots, CD_n$ represent the successive wave fronts after multiple reflections. For the nth beam the angle $D_n CB = 2n\theta$.

Consider the path differences of the first and nth beams at the point $P(x,y)$.

This is $\delta = PN - PM = PN - x$.

Now $$PN = x\cos 2n\theta + y\sin 2n\theta.$$

Thus $$\delta = x(\cos 2n\theta - 1) + y\sin 2n\theta.$$

Expanding the cosine and sine ($2n\theta$ being small) gives

$$\delta = 2nt\left(1 - \frac{2n^2+1}{3}\theta^2\right) - 2xn^2\theta^2.$$

By viewing fringes at the surface BC, x is made zero and the retardation lag behind the arithmetical value becomes

$$\frac{2n}{3}(2n^2+1)\theta^2 t,$$

which for the large values of n to be considered effectively equals $\frac{4}{3}n^3\theta^2t$, the approximate value stated previously. Brossell's method has considerable advantages in other connexions not to be pursued here.

It will be shown later that reflecting coefficients can be produced which are so high that 60–80 effective beams contribute to the interference effect. If we assume that the 60th beam, say, has fallen behind the first by $\lambda/2$, then beams of this order oppose the Airy summation and clearly reduce definition. Taking this as a convenient limit gives $\frac{4}{3}n^3\theta^2t = \lambda/2$. If X is the number of fringes per cm. on the wedge surface, $\theta = \lambda X/2$,

and therefore $$t = \frac{3}{2n^3\lambda X^2}.$$

For a wave-length in the green ($5{\cdot}5\times10^{-5}$ cm.) and $n = 60$,

$$t = \frac{1}{7{\cdot}92X^2}.$$

This value for t is the maximum permissible for close approximation to the Airy condition. In some instances fringes may be 1 cm. apart ($X = 1$). More frequently they are 1 mm. apart ($X = 10$) or even 0·1 mm. ($X = 100$). Table II gives the corresponding critical values of t.

TABLE II

Number of fringes per cm. X	1	10	100
Critical t, mm.	1·26	0·012	0·0001
d_{60} mm.†	0·25	0·025	0·0025

† See next section.

The calculation is not strictly applicable to the third case since t is less than a light-wave and the approximations are invalid.

However, an important feature emerges, namely that *the separation between the two surfaces must be as small as possible. This is the critical fact which has been overlooked by earlier workers.* When high magnifications are used it is essential to have fringes separated by less than a millimetre, for otherwise there are insufficient fringes in the field of view. It is clear that the separation between the surfaces must be reduced to the order of a *few wave-lengths of light at most*, otherwise definition suffers severely. This is one of the most important points in the technique.

Linear displacement of beams

The magnitude of the phase-lag effect depends upon the linear displacement of the successive beams along the surface. This is not only determined by the wedge angle θ, but also by the angle of incidence if this is other than normal. Clearly it is desirable to employ normal incidence unless the experiment forbids it.

The displacement of the beams is a matter of considerable importance when the surface under examination has a complex topography. It is essential to view interference from beams which have scanned as small an area as possible if confusion due to beams from different topographical features meeting at arbitrary points is not to occur.

The higher order beams come from regions progressively farther away from the first beam. To a first approximation the *linear* separation on the wedge surface between the first and nth beams is

$$d_n = 2n^2t\theta$$

$$= \frac{3}{2nX}$$

(by substituting $\theta = \frac{1}{2}\lambda X$ and $3/t = 2n^3\lambda X^2$).

For $n = 60$ this becomes

$$d_{60} = \frac{1}{4X} \text{ mm.}$$

The values of d_{60} for $X = 1, 10, 100$ are given in the third row of the previous table showing critical t values. It is seen from this that if t is of the order of $1/1000$ mm., as is desirable for the phase condition, then the 60th beam comes from a region only $1/400$ mm. away from the first beam. Thus all the relevant beams come from a region which is within the resolving limit of a typical low-power microscope objective. The interference pattern is then a correct picture of the topography. It will be noted that the value of t required is substantially the same as that necessitated by the phase condition.

Errors in collimation

Fabry [13] drew attention to the broadening effects due to lack of parallelism in the incident beam. This arises either through maladjustment of focus or through the finite size of the source stop. The broadening as a fraction of an order can be calculated as follows:

For normal incidence

$$n = 2t/\lambda.$$

For incidence at angle ϕ this diminishes to

$$n - \delta n = \frac{2t\cos\phi}{\lambda},$$

giving $$\delta n = \frac{2t}{\lambda}(1-\cos\phi) = \frac{2t}{\lambda}\,.\,2\sin^2\tfrac{1}{2}\phi.$$

Since only small values of ϕ are contemplated,

$$\phi = \sqrt{\frac{\lambda\,\delta n}{t}}.$$

The fringes usually have a half-width of about 1/40th order, and if the fringes are not to be increased in width by more than a fifth of this, $\delta n = 1/200$, giving for $\lambda = 5\times10^{-5}$ cm.,

$$\phi = \frac{5\times10^{-4}}{\sqrt{t}}\text{ radian} \doteq \frac{3\times10^{-2}}{\sqrt{t}}\text{ degree.}$$

Table III gives approximate values of ϕ and of the permissible stop diameters d for use with a 10 cm. focal length lens over the range of values of t.

TABLE III

t mm.	1	0·1	0·01	0·001
ϕ°	$\frac{1}{10}$	$\frac{1}{3}$	1	3
d mm.	0·2	0·6	2	6

With longer focal lengths, larger stops are permissible. Since the previous calculation shows that the typical case of fringes 1 mm. apart requires a value $t = 0{\cdot}01$ mm., such an arrangement can tolerate a 2 mm. source with a 10 cm. lens, with which collimation errors are not serious.

The experimental conditions for the production of highly sharpened multiple-beam Fizeau fringes are thus, that

(1) The surfaces must be coated with a highly reflecting film of minimal absorption.
(2) This film must contour the surface exactly and be highly uniform in thickness.
(3) Monochromatic light, or at most a few widely-spaced monochromatic wave-lengths should be used.
(4) The interfering surfaces must be separated by at most a few wave-lengths of light.
(5) A parallel beam should be used (within a 1°–3° tolerance).
(6) The incidence should preferably be normal.

In the following appendix the derivation of the Airy formula is given.

APPENDIX TO CHAPTER II

The Derivation of the Airy Formula

The following is a simplified derivation of the Airy summation formula. Consider the multiple beams brought to the focus of a lens, where the two silvered surfaces are parallel, distant t apart. It is assumed that there is no absorption. Let a plane wave front, of wave-length λ, fall on the interferometer, at an angle ϕ. Let T and R be the transmission and reflecting coefficients at each silver surface, the amplitudes of the respective beams being $T^{\frac{1}{2}}$ and $R^{\frac{1}{2}}$. Consider first the reflected light. Then the *amplitudes* of the successive multiple reflected beams which emerge and are collected by the lens are $R^{\frac{1}{2}}$, $R^{\frac{1}{2}}T$, $R^{\frac{3}{2}}T$ $R^{\frac{5}{2}}T$,..., etc. The emerging waves being parallel, sum at the lens focus. Each beam has a path difference $2\mu t\cos\phi$ behind its predecessor so that the phase difference between each beam is

$$\delta = \frac{2\pi}{\lambda}\,2\mu t\cos\phi.$$

Let the incident wave be represented by $\sin\omega\tau$ and the final result and by $D\sin(\omega\tau-\Delta)$, then we can write

$$D\sin(\omega\tau-\Delta) = R^{\frac{1}{2}}\sin\omega\tau+R^{\frac{1}{2}}T\sin(\omega\tau-\delta)+R^{\frac{3}{2}}T\sin(\omega\tau-2\delta)+\dots.$$

Expanding the sin terms and equating coefficients of sin and cos terms gives

$$D\cos\Delta = R^{\frac{1}{2}}+R^{\frac{1}{2}}T(\cos\delta+R\cos 2\delta+R^2\cos 3\delta+\dots),$$
$$D\sin\Delta = R^{\frac{1}{2}}T(\sin\delta+R\sin 2\delta+R^2\sin 3\delta+\dots).$$

These two combine in the usual manner to give

$$De^{i\Delta} = R^{\frac{1}{2}}+R^{\frac{1}{2}}T(e^{i\delta}+Re^{i2\delta}+R^2e^{i3\delta}+\dots)$$
$$= R^{\frac{1}{2}}+R^{\frac{1}{2}}T\,\frac{e^{i\delta}}{1-Re^{i\delta}}.$$

To simplify this, multiply and divide the imaginary part by $(1-Re^{-i\delta})$, making

$$De^{i\Delta} = R^{\frac{1}{2}}+\frac{R^{\frac{1}{2}}Te^{i\delta}(1-Re^{-i\delta})}{(1-Re^{i\delta})(1-Re^{-i\delta})}$$

$$= R^{\frac{1}{2}}+\frac{R^{\frac{1}{2}}T(\cos\delta+i\sin\delta-R)}{1-2R\cos\delta+R^2}.$$

The real and imaginary terms can be separated thus:

$$D\cos\Delta+iD\sin\Delta = R^{\frac{1}{2}}+\frac{R^{\frac{1}{2}}T(\cos\delta-R)}{1-2R\cos\delta+R^2}+\frac{iR^{\frac{1}{2}}T\sin\delta}{1-2R\cos\delta+R^2}.$$

Hence $$D\cos\Delta = R^{\frac{1}{2}}+\frac{R^{\frac{1}{2}}T(\cos\delta-R)}{1-2R\cos\delta+R^2}, \qquad (1)$$

$$D\sin\Delta = \frac{R^{\frac{1}{2}}T\sin\delta}{1-2R\cos\delta+R^2}. \qquad (2)$$

The intensity D^2 is the sum of the squares of (1) and (2), i.e.

$$D^2 = \left[R^{\frac{1}{2}}+\frac{R^{\frac{1}{2}}T(\cos\delta-R)}{1-2R\cos\delta+R^2}\right]^2+\left[\frac{R^{\frac{1}{2}}T\sin\delta}{1-2R\cos\delta+R^2}\right]^2$$

$$= \frac{4R\sin^2\frac{1}{2}\delta}{1-2R\cos\delta+R^2}$$

$$= \frac{4R\sin^2\frac{1}{2}\delta}{(1-R)^2+4R\sin^2\frac{1}{2}\delta}.$$

This represents Airy's formula for the *reflected* intensity and is a quantity that will be required later. Let B^2 be the *transmitted* intensity, then

$$B^2 = 1-D^2$$

$$= 1-\frac{4R\sin^2\frac{1}{2}\delta}{1-2R\cos\delta+R^2} = \frac{(1-R)^2}{1-2R\cos\delta+R^2}$$

$$= \frac{(1-R)^2}{(1-R)^2+4R\sin^2\frac{1}{2}\delta}.$$

But $1-R = T$ if there is *no* absorption,

hence $$B^2 = \frac{T^2}{(1-R)^2+4R\sin^2\frac{1}{2}\delta},$$

i.e. $$B^2 = \frac{T^2}{(1-R)^2}\left(\frac{1}{1+\{4R/(1-R)^2\}\sin^2\frac{1}{2}\delta}\right),$$

which is Airy's formula for the *transmitted* intensity.

When $\sin^2\frac{1}{2}\delta = 0$ this is a maximum and has the value $T^2/(1-R)^2$, which is unity since $T = 1-R$, hence *without absorption* the transmitted fringe maxima have intensity equal to that of the incident light, *no matter what the reflecting coefficient may be.* This occurs when

$$2\mu t\cos\phi = n\lambda.$$

The minima occur at $2\mu t\cos\phi = (n+\frac{1}{2})\lambda$, i.e. when $\sin^2\frac{1}{2}\delta = 1$. This makes $B^2 = T^2/(1+R)^2$, which can also be written $B^2 = \{(1-R)/(1+R)\}^2$. Clearly when R is high the minima are very weak.

The following graphical method (shown to the writer by R. G. Lunnon in 1927) is a simple elegant method for deriving the shape of the transmitted fringe distribution given by the Airy formula (reflected fringes, being complementary, are obtained by inverting the curve obtained).

Since $$B^2 = \frac{(1-R)^2}{1-2R\cos\delta+R^2},$$

construct a triangle within a circle of radius unity as shown in Fig. 12 *a* in which $XY = 1$. Let the distance XZ be equal to the reflection coefficient R and let angle $YXZ = \delta$. Then $ZY^2 = 1-2R\cos\delta+R^2$.

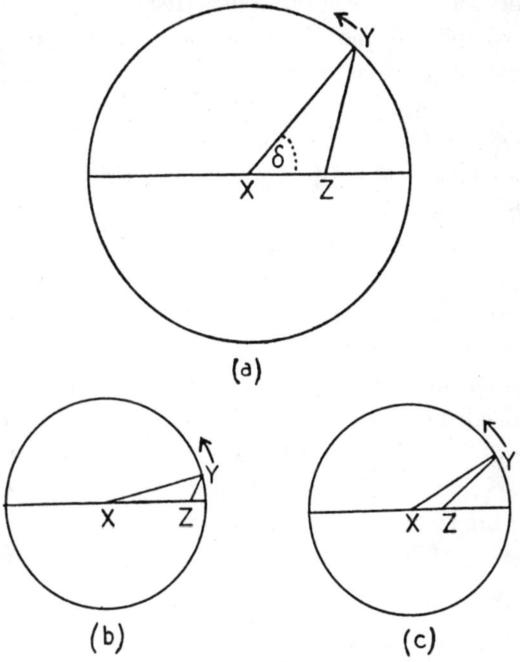

FIG. 12.

Thus clearly as Y sweeps round the circle (i.e. as δ goes through to complete an order) then $1/ZY^2$ will be a measure of the variation of B^2 with δ. Hence a plot of $1/ZY^2$ against δ will give the fringe shape. The absolute intensity values can readily be obtained by multiplying the ordinates by $(1-R)^2$.

It can easily be seen from this method how the fringe shape depends on R. Take the case where R approaches unity, as illustrated at (*b*). For small δ value (near $n\pi$) YZ is quite small so that $1/YZ^2$ is large, but as can be seen YZ increases at a rapid rate as Y sweeps round, hence the intensity drops rapidly.

But in Fig. 12 *c*, where R is small, the quantity YZ is initially large and only increases slowly as Y moves round, hence broad maxima result.

CHAPTER III

THE SILVERING TECHNIQUE

Introduction

THE writer has given elsewhere [14] a detailed account of the techniques used for the preparation of mirrors for Fabry-Perot interferometers. The technique described here is similar, but there are simplifications. Two major problems in using the Fabry-Perot interferometer are fortunately avoided. These are (1) source-intensity difficulties arising when weak spectrum lines are being studied, a factor which often severely restricts the permissible thickness of silver and limits the fringe sharpness obtainable; (2) in spectroscopic Fabry-Perot studies it is frequently necessary to cover a wide wave-length range and considerable difficulty is experienced in finding mirrors which have high reflection (and also low absorption) over a wide colour range.

Fortunately in topographical studies precision fringes are generally needed only over the range between the red and green (say 7,000–5,000 A.U.), and in this region it is possible to obtain and use reflecting coefficients exceeding 0·90 by using evaporated silver. For interferometry, evaporated silver is superior to sputtered silver, so that only the silver evaporation technique will be described.

Surface cleaning

When a reflecting film is to be evaporated on to a surface a condition of cleanliness is of paramount importance. Oil and impurity films reduce the reflecting coefficient somewhat and increase the absorption a great deal, both of which are undesirable. The cleaning treatment to be used depends upon the nature of the surface about to receive the silver. Typical surfaces will be considered in turn.

(*a*) *Optical flats of glass or quartz*

A preliminary washing with soap and water will remove gross amounts of grease contamination from handling. With costly accurately worked flats, the usual process of cleansing with concentrated acids and alkalis should be avoided. In any case quartz is soluble in caustic soda. Tenacious oil films can be removed in the simple degreasing plant described later, but this is rarely necessary. A suitable chemically mild efficient cleansing agent is strong H_2O_2, which removes many organic materials and is also an excellent solvent for removing silver from a surface that requires recoating. Cleansing is carried out by rubbing gently (cold) with damped pieces of well-boiled linen and then polishing with a dry specimen of the same material.

The 'breathing test' is a good index of the degree of cleanliness achieved. By gently breathing on the surface, moisture condenses into a film of minute droplets and this shows a 'figure' revealing imperfectly polished areas. An oil-free surface condenses moisture in an invisible film. Even a monomolecular grease film breaks this up into droplets and the surface appears grey matt. This is the normal condition after cleaning. Rayleigh showed that a piece of glass passed rapidly through a flame becomes film-free, and preheating a polished surface to 250° C. has a similar effect. It has been shown that the cleansing mechanism of the flame is due to thermal ions and that this same cleansing effect is obtainable in a vacuum by means of a glow discharge, in a manner to be described later.

(*b*) *Cleavage surfaces*

Freshly cleaved crystal surfaces (e.g. mica, selenite, calcite) are quite clean and require no treatment other than degassing in a vacuum, and possibly gentle ion bombardment.

(*c*) *Resistant surfaces*

Resistant surfaces (e.g. diamond, polished or natural) can be cleaned with fairly severe reagents. Ion bombardment is used as the final treatment.

(*d*) *Metals*

Unless etching is required, metal surfaces must be carefully treated. After polishing, a safe procedure is to degrease with ethylene trichloride. The ethylene trichloride is gently boiled in a beaker closed with a watch-glass. The metal is suspended above the liquid and the vapour condensing on it removes the greases efficiently, the contaminated liquid dropping down into the beaker. Ionic bombardment is the final agent.

The evaporation technique

The deposition by evaporation of mirrors suitable for multiple beam interferometry was developed first by Ritschl [15]. The

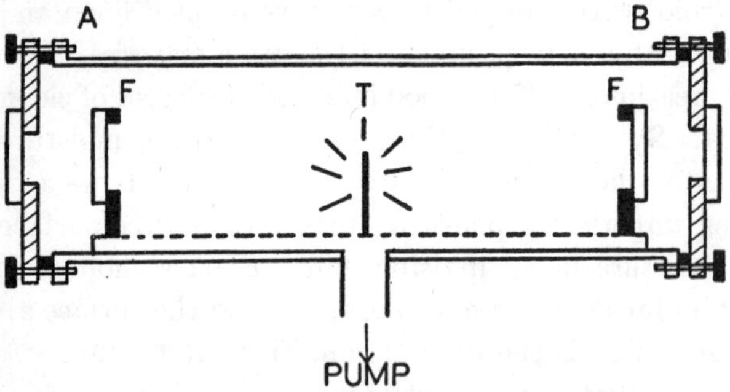

FIG. 13.

two methods described below are improvements on this procedure. In both cases the silver is evaporated from a suitable filament in a high vacuum (preferably at least 10^{-5} mm. of Hg). Two evaporating plants have been used, a horizontal and a vertical. With the horizontal plant the two mirrors required can be made simultaneously, whereas with the vertical arrangement, materials other than silver can be deposited (e.g. the cryolite used for specific purposes to be described later). The horizontal plant is shown in Fig. 13.

A metal cylinder 80 cm. long and 8 cm. in diameter is closed at both ends, the joints being made vacuum tight with rubber gaskets. Insulated metal rods pass through the metal end-pieces, which are also provided with windows. A pumping system, such as a Metrovac 02, 03 combination diffusion pump, is used for evacuation.

The surfaces to be coated, *FF*, are mounted on a framework and at the centre of this is a filament which carries the silver. The filament developed and described here has been found satisfactory. Two variants of a suitable form of filament are shown in Figs. 14*a* and 14*b*. A strip of molybdenum, 3 mm. wide, is bent into a loop or several loops of diameter some 4 mm. A piece of silver is placed within the loop, practically filling it. On passage of current the silver melts and owing to its high surface-tension fills the loop on both sides with the result that both

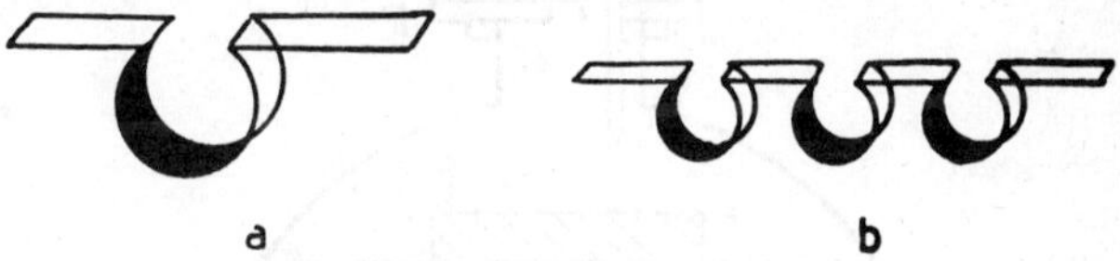

FIG. 14.

receiving surfaces are faced by a vertical relatively large disk of molten silver, and deposition takes place rapidly, despite the long distance. Since only the *edge* of the filament strip faces the receiving surfaces, the ratio of the area of silver to that of molybdenum is considerable. This reduces the evaporated molybdenum to negligible proportions.

The technique of deposition is as follows. The system is pumped to about 0·1 mm. pressure with a rotary pump, at which stage the silver is pre-fused, melts, and degasses. The mean free path is so small that no silver can reach the surfaces. A high-voltage glow, provided by a large spark coil, is then passed between the electrodes. This is important, for it provides the ions for cleaning by ionic bombardment. After some 15 minutes the diffusion pumps are set in operation, and when at least 10^{-5} mm. pressure is reached evaporation is begun. It takes only 1 to 2 minutes to deposit the amount of silver required. This is estimated by viewing a bright lamp through the windows and evaporating to a standard of transmission, either visually or by photocell. The final colour and intensity (a deep blue) can only be learnt by experience. Purity conditions are important.

Uniformity of the silver deposit

The uniformity of the silver deposit is of paramount importance. The amount of condensation, in a perfect vacuum from a point source, at any point on a plane surface is inversely proportional to the square of the distance of the point from the source and to the cosine of the angle between the normal to the plane and the line joining the point in question to the source. If P is the density at a point for which this angle is ψ, then

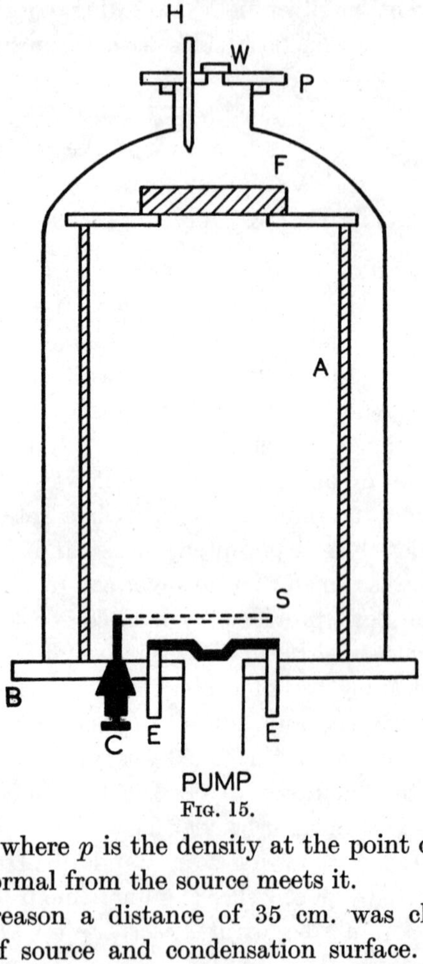

FIG. 15.

$P = p \cos^3\psi$ where p is the density at the point on the surface where the normal from the source meets it.

For this reason a distance of 35 cm. was chosen for the separation of source and condensation surface. Suppose the surface to be examined is a disk of 1 cm. radius, then $\cos^3\psi$ for the circumference is 0·9988.

The density at the circumference is therefore less than that at the centre by only 0·12 per cent. Clearly over most of the disk the silver can be considered uniform, even with a point source.

However, even this is much improved upon by the employment of an *extended disk* source. The most stringent interference tests fail to reveal any serious variations in average silver thickness over such a surface, whilst no local variations can be detected at magnifications of $\times 400$.

The vertical evaporator is shown schematically in Fig. 15. A tall narrow bell jar, over 35 cm. long, stands upon a phosphor-bronze base, the junction being vacuum tight. The jar sits in an annular recess upon a gasket of neoprene rubber, and the base sits upon an oil-diffusion pump, the opening of which has a baffle protector. Insulated water-cooled electrodes *EE* pass through the base and carry a thick molybdenum heating-strip. There are several electrodes linked up to three separate molybdenum strips, each of which can carry its own metal charge. By means of the vacuum-tight cone *C* a shutter *S* can be swung in and out of position over the filament. The surface *F*, receiving the silver, rests upon a glass tripod *A*. The top of the bell jar can be sealed either with a rotating cone to permit movement of *F* or alternatively with a metal plate *P* which has in it a small observation window *W* and an electrode *H*, used for the degassing discharge.

Before deposition is begun the filament is heated with shutter *S* in position over the strip. Any impurities are burnt off and thus prevented from reaching *F*. Such impurities can have a serious influence in increasing film absorption. The shutter is removed after heating the strip for half a minute and deposition then begun.

The reflectivity of silver films

With careful attention to purity, it is possible to obtain with silver a reflection coefficient of 0·94 in the green and a reasonable transmission (1 per cent.). No other metal is so efficient (in theory one might expect sodium to be better, but a sodium film is not a practical possibility). It is quite unnecessary to evaluate numerically the actual reflectivity, since what is required is the highest reflectivity consistent with the thickness permitted by the light source available.

It will be shown later that for topographical interferometry the light sources now available are much brighter than those normally used in hyperfine structure studies with the Fabry-Perot instrument. The result of this is that thicker silverings are permissible, with consequent higher reflectivity and therefore sharper fringes. The reflectivity R, transmission T, and

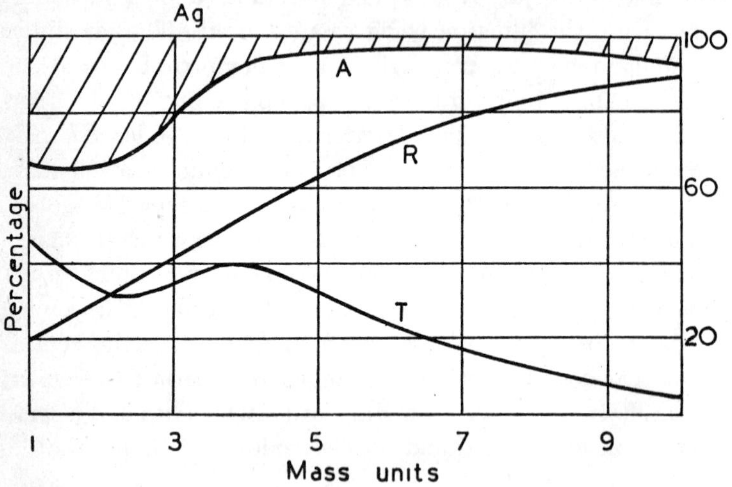

FIG. 16.

absorption A depend upon film thickness for green light in the manner shown in Fig. 16.

At a thickness of about 500 A.U. the reflectivity reaches 0·90. The absorption, shown by the shaded area, is large for low reflectivities, reaches a minimum when R is 0·75, and then increases slowly. An increase in thickness beyond 500 A.U. produces a small increase in reflectivity but a relatively considerable increase in absorption.

The influence of absorption on transmitted intensity is shown by Fig. 17, in which $\{T/(T+A)\}^2$ is plotted against reflectivity. The values used for the calculation have been determined in the writer's laboratory by W. K. Donaldson. The rapid fall in intensity with increase in R is clearly shown and at $R = 0{\cdot}94$ only 1 per cent. of the light is effective. With badly prepared silver films the transmission may be much less.

A film thickness giving $R = 0{\cdot}94$ is clearly prohibitive unless very intense sources are available. This is fortunately the case.

For the estimation of the film efficiency an empirical rapid method involving both reflectivity and absorption is available.

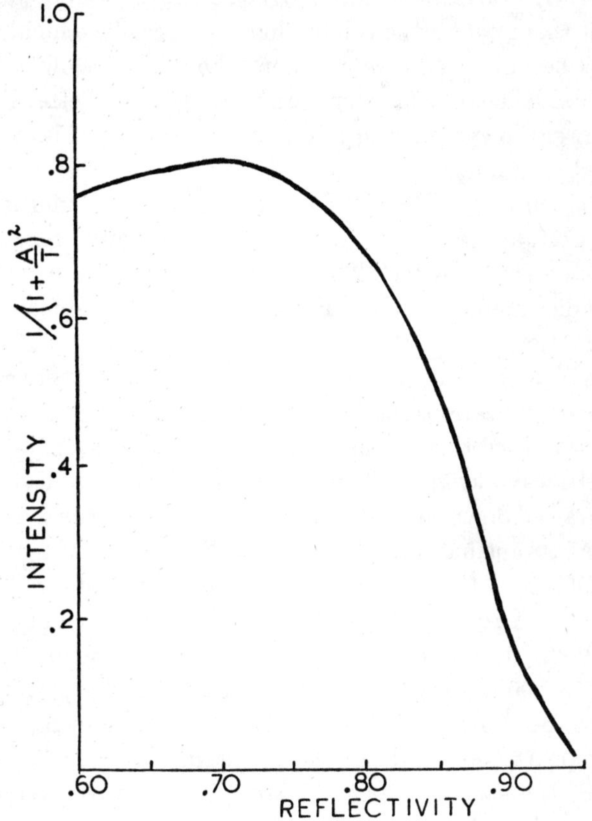

FIG. 17.

Silver is deposited on to two glass disks, some 5 cm. in diameter, and these are held close together, almost parallel, and close to the eye. A bright filament lamp distant about 1 metre is viewed through them and by slightly tilting one plate a succession of multiple images can be seen. The colour characteristics and the count of images suffice to indicate the quality and usefulness of the silvering. If the incident beam is supposed to have unit intensity, the nth image has intensity T^2R^{2n} and unless R is high the intensities fall off rapidly. With high absorption T is low and the whole series is weakened.

A good pair of mirrors with transmission of a few per cent. and reflectivity of over 0·90 shows over 80 clearly defined images.

A further guide is the colour characteristic. The higher-order images become progressively more and more yellow since the longer wave-lengths have higher reflectivity. The better the film, the slower the approach to yellow and the more blue is to be seen in the initial images.

Although the thickness of a satisfactory silver film is about 500 A.U., it appears to contour the surface structure very closely, at least to within 10 A.U. This will be proved later when discussing studies made on mica. Electron microscopy offers further confirmation.

It will be shown later that white-light fringes have many valuable applications, and it is found that the maximum reflectivity (for very thick films, equivalent to solid silver) depends upon the wave-length. There is a regular fall in reflectivity from the infra-red down towards the blue. At λ 10,000 the reflectivity is some 0·98 and has fallen to 0·95 at λ 6,000. At λ 5,000 it is 0·93, at λ 4,000 it is 0·90, from which point on it drops with great rapidity to about 0·04 (i.e. like glass) in the neighbourhood of λ 3,000. It then rises again. In the visible region between λ 4,000–λ 7,000 it is possible to use silver to give sharp white-light multiple-beam fringes. It is usually advisable to operate near to the region λ 5,000, for the reflectivity is high here and the dispersions of the spectrographs employed are usually good in this region. Fringe definition with red light is much superior to that with blue light, but this is only partly due to the silvering. Prism spectrographs have a higher dispersion in the blue than in the red, and this makes blue fringes appear relatively broader than red fringes. One frequently aims at as high a fringe dispersion as possible, i.e. the fringes are spread over as wide a wave-length range as can be tolerated without excessive broadening. In applying such a procedure it *must be remembered that fringe definition always falls steadily towards the blue end of the spectrum.*

It is only possible to obtain reflectivities in the blue approaching 0·90 if very dense silver films are used. Under such circumstances the transmission throughout the spectrum is usually too low for practical application. The blue fringes alone can, however, frequently still be photographed because of the much greater sensitivity of photographic plates for blue light.

CHAPTER IV

SILVER-MODIFIED NEWTON'S RINGS

Light sources

THE modified Newton's rings formed when a suitable silver film is deposited upon the two opposing surfaces illustrate a number of characteristic properties of multiple-beam Fizeau fringes and will be discussed for this reason [17].

The most convenient source for Fizeau fringe interferometry is a mercury lamp. For such a source efficient filters are available which pass either (*a*) the green line, λ 5,461, or (*b*) the yellow lines λ 5,770, 5,790, or (*c*) the green and yellow lines together. There are three convenient mercury sources: (*a*) a vacuum arc; (*b*) a high-pressure arc of the 'Osira' type; (*c*) a high-pressure 'point-source' arc of the 'Sieray' type. The line width in the vacuum arc is due largely to hyperfine structure and is less than an Ångstrom unit. The line width in the 'Osira' type arc when run with normal current is sufficiently small to be of no consequence *provided that the separation between the interference surfaces is less than* 0·01 mm. With the high-pressure 'Sieray' point source the line width increases with the operating current and may exceed 5–10 A.U. Even with low currents, however, this source is so brilliant that thick silver films can be tolerated and thus reflecting coefficients of at least 0·90 employed.

Newton's rings

The multiple-beam Newton's rings given by the green line are shown in Fig. 18 with the optical arrangement shown in Fig. 19; the light is incident normally, the source being at *A* and the fringes at *B*. The inherent surface defects on the glass surfaces are rendered visible by the uneven character of the much sharpened rings. The change in level on moving from ring to ring is $\frac{1}{2}\lambda$. It is clear that irregularities less than 1/150th of an order can certainly be measured and this quantity, less than 20 A.U., is of molecular dimensions.

For any given ring, of order n, we have $n\lambda = 2t$ (if $\mu = 1$), so that $dt = dn . \frac{1}{2}\lambda$. To evaluate a small change in thickness dt, it is only necessary to determine the fraction of an order, dn, produced by that change.

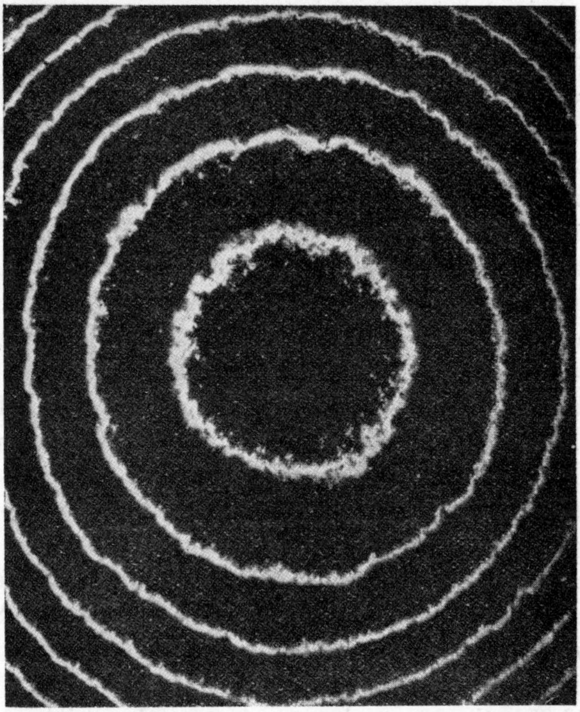

FIG. 18.

It is to be noted that these rings are *transmission* fringes. If viewed in *reflection*, with the arrangement of Fig. 1, the

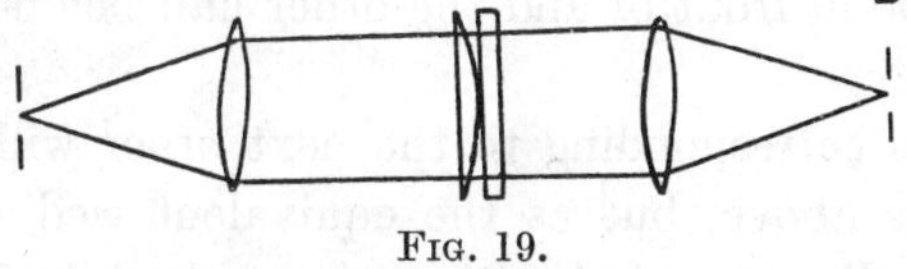

FIG. 19.

complementary system, fine dark lines on a bright background, is seen. Such a system would also arise if a silvered lens rested upon an opaque metal reflecting surface.

Non-normal incidence

Interesting new interference phenomena appear when the parallel beam of light is incident at angles other than the normal. These phenomena have a bearing on the interferometric practice to be adopted generally and will therefore be discussed in some detail.

If the lens-plate combination is tilted through an angle ϕ it follows from the geometry that the circular fringes will become elliptical. Furthermore it can be shown either from simple geometrical construction or more elaborately, as Feussner [18] has done, that a change in location of the fringes takes place. Observation and theory both show that the transmitted elliptical rings lie on regular curves, one half of the system before and the other half behind the interference film. As a result of this peculiar focal location, a photographic plate set normal to the light beam records only a few fringes simultaneously in focus.

To a first approximation the fringes can be considered as wedge fringes, since each fringe is localized and arises from multiple reflections over a small area. The effective wedge-angle is not constant but increases progressively with fringe order number. It can be considered that at any point the interference is that of a wedge, the angle of which is the angle made by the tangent to the lens, at the point in question, with the plane glass surface.

Considering the Newton's rings (or, for simplicity, a single ring only) as arising from a double wedge, it is clear from the construction shown in Fig. 20 that at non-normal incidence the wedge AOB leads to a fringe at the point X, whilst the wedge COD leads to a fringe located at Y. Hence one-half of the ring system will be in front of and the other half behind the interference film.

The fringes corresponding to the next order will lie on the same sides as above, but as the equivalent wedge-angle has changed, the distance of the fringe from the interference film changes. The following approximate theory of the fringe location is in qualitative agreement with the observed positions.

Feussner has shown that the fringes formed by multiple reflection of light incident at an angle ϕ on to a wedge of thickness d and wedge-angle θ, with refractive index μ, appear at a distance D from the wedge given by

$$D = \frac{\sin\phi\cos^2\phi}{\mu^2 - \sin^2\phi}\frac{d}{\theta}.$$

For an air film this reduces to $D = (d/\theta)\sin\phi$. It is legitimate to

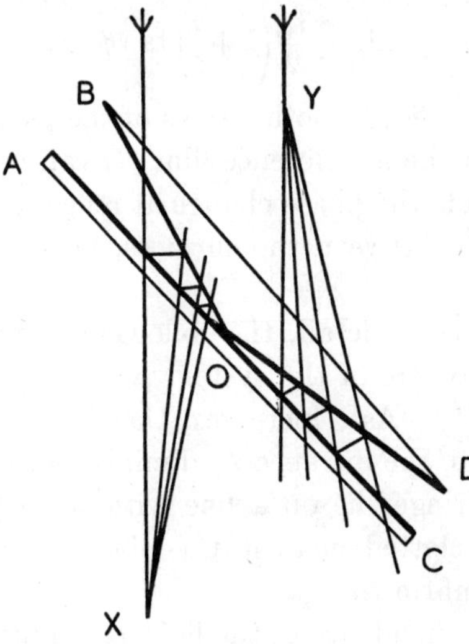

FIG. 20.

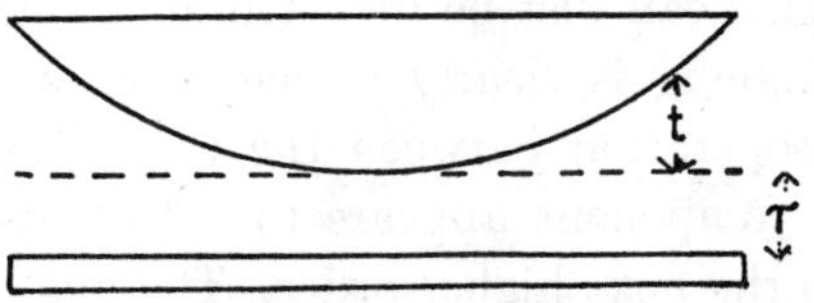

FIG. 21.

apply this to the Newton's ring fringes, using for each fringe the film thickness as d, and the angle between the plane and the tangent as θ.

For the nth ring the wedge thickness is $t+\tau$ (see Fig. 21), where $n\lambda = 2(t+\tau)\cos\phi$, and τ accounts for possible imperfect contact between the glass surfaces. The angle θ between the tangent and the plane is given by $\sin\theta = \theta = \rho_n/R = 2t/\rho_n$, where ρ_n is the radius of the nth ring. The distance D_n of the nth ring from the interference film is thus

$$D_n = \sin\phi \frac{t+\tau}{2t} \rho_n.$$

Since the fringes are viewed in the direction of incidence, they appear as ellipses, and the measured radius of the smaller axis r_n is $\rho_n \cos\phi$, so that

$$D_n = \frac{r_n}{2}\left(1+\frac{\tau}{t}\right) \tan\phi.$$

The argument applies to both halves of the pattern, located on opposite sides of the interference film. (τ can be taken for this analysis to include the phase change at reflection as well as the metrical distance between the surfaces, i.e. it is the 'optical' distance.)

Examining this in detail, the following conclusions can be drawn. At the centre of the system (where $t = 0$) the fringes will be at infinity. As t increases, that is, as the ring order number increases, the distance D_n diminishes with rapidity and ultimately the fringes lie on a line with slope $(\tan\phi)/2$ if the distance D_n is plotted against the observed ring radius r_n. Observations confirm this.

As the angle of incidence of the light is gradually increased a new effect, namely doubling of the fringes, begins to make its appearance. This can just be detected at an incidence of 20°. At 30° the doubling is clearly marked, and as the incidence increases the separation between the two components grows steadily. One component appears to detach itself and move across towards the next higher order. The march of the outer component with increasing incidence is shown clearly in Fig. 22.

Attention may be drawn to three special characteristics. As the incidence increases

(*a*) the doublet separation grows regularly;

(*b*) when first resolved, both components have the same intensities, but as the outer higher-order fringe moves away it becomes progressively weaker, finally vanishing as it approaches the next order;

(c) as the incidence increases, the outer component becomes progressively sharper and sharper, being extremely narrow at the higher angles of incidence. (This is to be observed best in the small region where the focus is correct.)

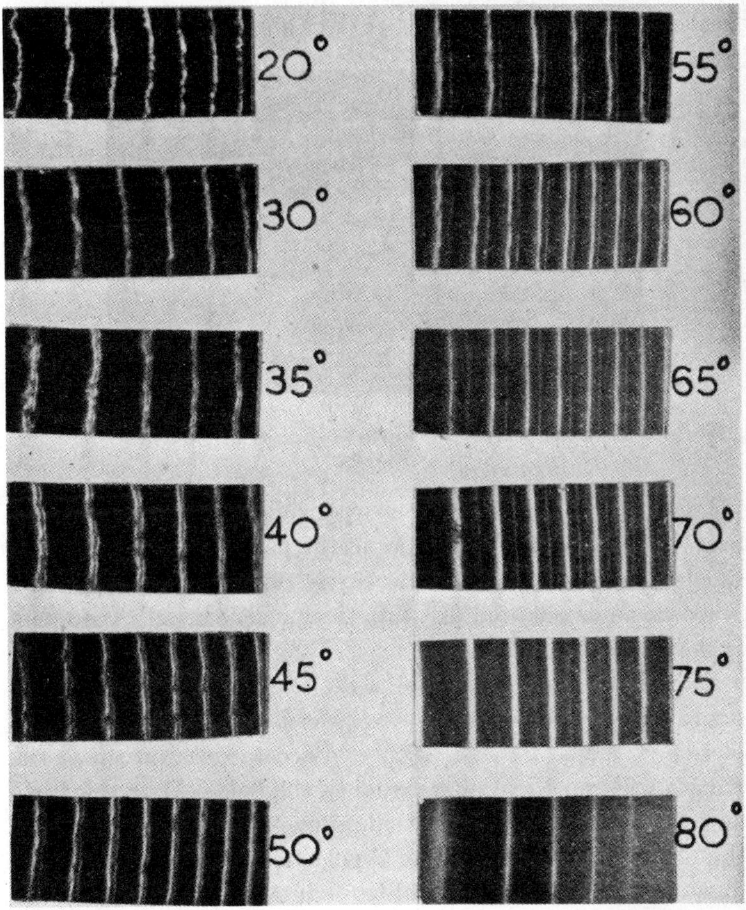

FIG. 22.

Although at normal incidence the fringes are quite sharp and narrow, the sharpness of both members of the doublet increases still farther with incidence. The stronger fringes sharpen up much more slowly than the weaker outer systems. This difference in fringe width can only be interpreted as meaning that the reflection coefficients for the two sets are increasing with incidence at different rates, for the fringe width is sensitive to the value of the reflecting coefficient.

This observation suggests that the two sets of rings consist respectively of light polarized at right angles and parallel to the plane of incidence, the outer sharper rings being formed by the beam which has the magnetic vector parallel to the plane of incidence. This is proved to be correct by the introduction of a Nicol prism or polaroid disk into the incident beam. The effect is shown clearly in Fig. 23.

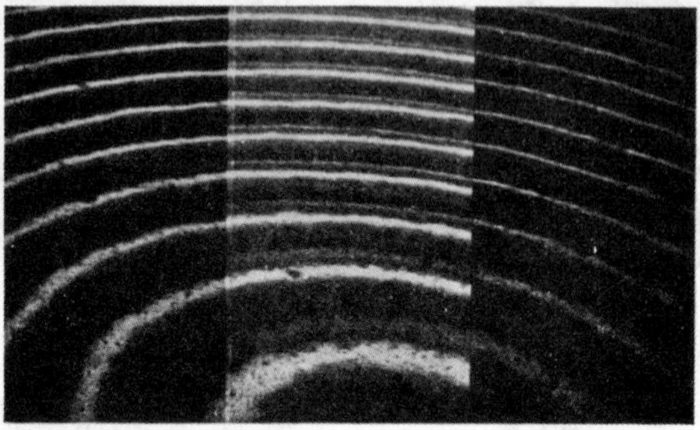

(*C*) (*B*) (*A*)

FIG. 23.

This shows the appearance of the fringes with angle of incidence 55°. A triple shutter was placed over the photographic plate and the sections *A*, *B*, *C* exposed in turn. *B* shows the fringes with no Nicol or polaroid in the field. At *A* the Nicol or polaroid is set to pass the vibrations which are polarized in the plane of incidence and at *C* polarized perpendicular to the plane of incidence. If a Nicol is used, care must be taken not to alter the angle of incidence through the deviation of the prism. A polaroid disk obviates this difficulty. The exposures have been adjusted to bring out the effect of the polarizer, and the weaker parallel vector component, at *A*, has been given twice the exposure of the stronger at *C*, the correct intensity ratio being seen at *B*.

The origin of the doubling of the rings

The doubling of the fringes arises because of the phase change taking place at reflection at a metallic surface. It is well known from classical electromagnetic theory that in the case of non-normal incidence the phase change is different for parallel and perpendicularly polarized light. At normal incidence the differential phase change is zero. As the incidence increases the differential phase change grows, and this effectively alters the optical path between the metal surfaces.

In general, as represented in Fig. 21, the lens is not in perfect contact with the glass plate. Let the two glass components be separated by a distance τ in a medium of unit refractive index, and let ϵ be the effective path change at a single reflection. Then the path difference between the succeeding beams, which add up to form the series due to multiple reflections, is

$$2t\cos\phi+2\tau\cos\phi+2\epsilon = n\lambda,$$

in which ϕ is the angle of incidence. If ρ_n is the radius of the nth ring, then

$$\rho_n^2 = \frac{R}{\cos\phi}(n\lambda-2\tau\cos\phi-2\epsilon),$$

from which it follows that an increase in the ring diameter arises from a negative phase change, i.e. one which effectively reduces the optical path.

On differentiating, $d\epsilon = \frac{1}{2}\lambda\, dn$, giving the differential phase change $(d\epsilon/\lambda)$ as $\frac{1}{2}dn$, which is half of the observed fraction of an order separation between the two fringe components. This is independent of the ring order number, the radius of curvature, and does not explicitly involve ϕ although it is a function of the angle of incidence.

The differential phase change at reflection can thus, for the first time, be directly and objectively determined from a fringe displacement. In the former, now classical, determinations of this quantity, the phase shift was not directly observable but was derived from measurements of the degree of elliptical polarization of the light reflected from the metallic film. The fringe-doubling appears in both the transmitted and the reflected systems and the behaviour is identical. A striking feature in both is the rapid falling off in intensity of the fringes due to the beams polarized with the magnetic vector in the plane of incidence. This is evidence of a differential absorption coefficient.

The evaluation of *dn*

The accurate determination of the order-change *dn* is one of the principal sources of error in Fizeau fringe interferometry. In general the surface examined has an irregular topography, as a result of which the Fizeau fringes have a variable dispersion. If this dispersion varies rapidly, accuracy inevitably suffers. If, on the other hand, a region of reasonably constant, or alternatively, uniformly and slowly varying, dispersion can be selected, then precision is attainable by adopting the method of approximation originally proposed by McNair [19] for the Lummer plate interferometer.

Let $A, A', A'', A''',...$ represent the positions of a series of fringes, exhibiting a regular change in dispersion due to a curved surface, and let B, B', B'', $B''',...$ be the second subsidiary system. To a close approximation the fraction of an order separation *dn* is given by

$$dn = \frac{A'B'}{\frac{1}{2}(BB'+A'A'')} = \frac{A''B''}{\frac{1}{2}(B'B''+A''A''')} = \text{etc.}$$

A local irregularity can seriously disturb these relationships and it will be shown later how this difficulty is overcome by using different fringe types. It is implicitly assumed here that *dn* is less than one and also that the correct allocation to orders is known. Thus it is taken for granted that the fringe separation is $A'B''$ and not $B'A''$. In this particular example there is no ambiguity since the increase of fringe separation with angle of incidence can be tracked through from zero separation at normal incidence. This is a rare and unusual case and special techniques are often required to allocate the orders correctly.

The differential phase change

The differential phase change depends only slightly upon the thickness of the silver film. The fringes reproduced in Fig. 23 were obtained with a silver film of thickness about $\frac{1}{10}\lambda$, i.e.

about 550 A.U. The way in which this phase change depends upon the angle of incidence is shown in Fig. 24, in which angle of incidence is plotted against the phase, given as a fraction of a wave. The experimental curve is I. A theoretical curve II is that given by MacLaurin [29]. The curve III shows the difference between these.

Up to 60° incidence the experimental and theoretical curves

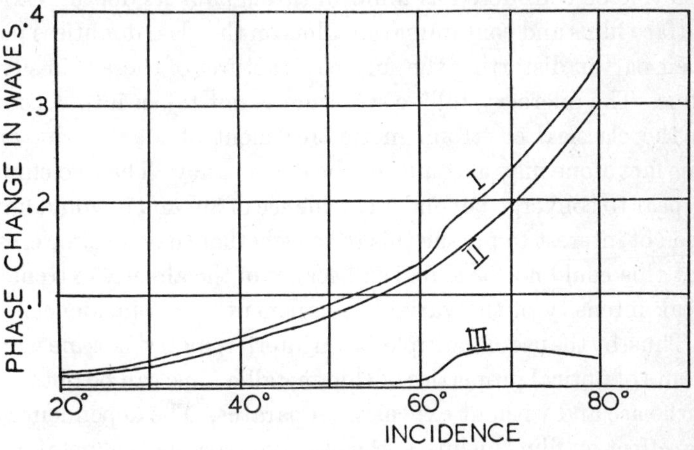

FIG. 24.

run closely parallel, the theoretical being systematically lower than the experimental by the small amount of 0·007. This difference is undoubtedly real, for the experimental precision is high and it can be shown that any uncertainty in the film thickness has a much smaller effect.

Beyond 60° there is a more marked departure from theory. The experimental curve shows a point of inflexion, the position of which was found with accuracy by increasing the number of observations in this neighbourhood to one per degree. After inflecting, the experimental curve converges slowly on the theoretical one as the angle of incidence increases to high values.

The difference between the curves I and II is plotted below as curve III. This difference curve cannot be attributed to any false estimate of film thickness and represents a real discrepancy between theory and observation. Thus, whilst to the first order theory is vindicated, there exists a second-order difference, the main characteristic being the point of inflexion which occurs in the neighbourhood of 63°.

A possible source of the discrepancy is the effect of transition at the surfaces. The refractive index does not change abruptly at the front or the back surface of the silver film. A film $0{\cdot}1\lambda$ thick is perhaps little more than 200 atoms deep, and in no sense can this be considered as uniform throughout its depth. Oxide surface films and contamination films on the glass doubtless play their part in distorting the surface structures of these thin silver films. The microcrystalline structure is not taken into account in the classical electromagnetic treatment of MacLaurin, and this fact alone may account for the discrepancy. The two curves appear to converge beyond an incidence of 80° and it would have been of interest to pursue this to see whether they meet or cross, but this could not be achieved because of the already extremely weak intensity of the variable component at 80° incidence.

Thus, by the use of multiple-beam interferometry, new measurements of optical properties of thin metallic films can be obtained with ease and without expensive apparatus. The dependence of the effect on film thickness, the effect of varying the metal, and the nature of the interface material (e.g. liquids can be introduced between the surfaces), can all be readily investigated.

It is clear both from theory and observation that the differential phase displacement is strictly zero only at normal incidence. Below 20° the doubling is not resolved but the effect appears as a fringe broadening. It has already been shown that for collimation purposes angles up to 3° can be tolerated, and with such angles the broadening due to differential phase change can be disregarded.

The doubling due to differential phase change appears in some of the experiments to be described later.

CHAPTER V

CRYSTAL TOPOGRAPHY USING FIZEAU FRINGES

QUARTZ, MICA, SELENITE, CALCITE

Introduction

THE use of *two-beam* interference methods for the examination of crystal surfaces is a fairly old technique. Thin mica slips for use as windows in alpha-particle experiments were usually selected by this procedure. Siegbahn [21] reported on the examination of crystal surfaces by this means in 1933, and more recently Kayser [22] has described observations made on diamond surfaces. Buckley [23] also noted the formation of Fizeau fringes in thin plates of potassium chromate (1932).

As already emphasized, however, the fringe-intensity distribution with only two beams is quite inadequate for the resolution of fine detail. The first application of *multiple-beam* methods to the examination of a natural crystal face was made with a highly lustrous (100) face of a left-handed quartz crystal [23]. The details will now be given since a number of general principles are involved.

Quartz

The particular crystal examined possessed a (100) face with a high natural polish, about 1 sq. cm. in area, and is illustrated face on and in profile in Fig. 25. Examination with a hand lens, using critical illumination, revealed a number of surface markings included in Fig. 25*a*. Since it is necessary to produce interference at normal incidence, the incident parallel beam must traverse the path *PQR* shown in Fig. 25*b*. This required the light to pass through the severely striated vertical pyramid face, which largely destroyed effective collimation with the result that the fringes were considerably broadened. This serious defect was remedied by contacting a thin microscope slide at *Q*, on to the striated surface, using Canada balsam as a medium. This treatment effectively destroyed the striae, and correct collimation was restored.

The (100) surface was coated with a suitable thickness of silver and then mounted against a similarly silvered optical flat.

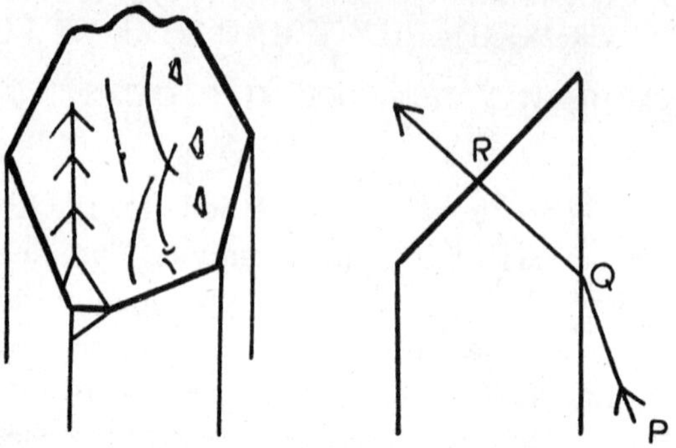

(*a*) (*b*)

FIG. 25.

The optical flat

For topographical studies the optical flat used must be of very high quality. The types often used for high-grade Fabry-Perot, interferometers are thick disks of quartz, fused silica, or glass, usually flat over a large area (5 cm. disk) to within $\lambda/60$ at least.† *Locally the error is much less* as revealed by observations described later.

The higher the magnification used, the less important is general curvature, but the more important are local small-scale errors. Experience shows that a flat made of polished glass has an extremely good local structure. It appears that the glass flows during polishing, leaving a surface smooth locally to within molecular dimensions apart from characteristic polish scratches. In all topographical studies using an optical flat there is a residual very small fine-grain fringe structure equivalent to effective height changes of molecular dimensions, and this is due to fine polish marks on the flat. Apart from such fine-grain detail, a ready check on topography is available by simply matching the crystal against different portions of the flat. Hilgers have made for the writer a group of $1\frac{1}{2}$ in. flats, 3 mm. thick, of which the central $\frac{1}{2}$ in. disk has been worked flat to better than $\lambda/40$. These flats are locally, over appreciable areas, certainly smooth and flat to within at least $\lambda/200$, probably much better.

† Flat errors are usually given in terms of the green mercury line.

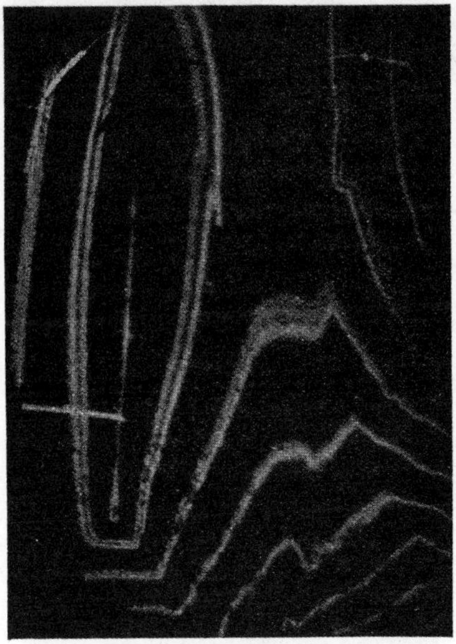

FIG. 26.

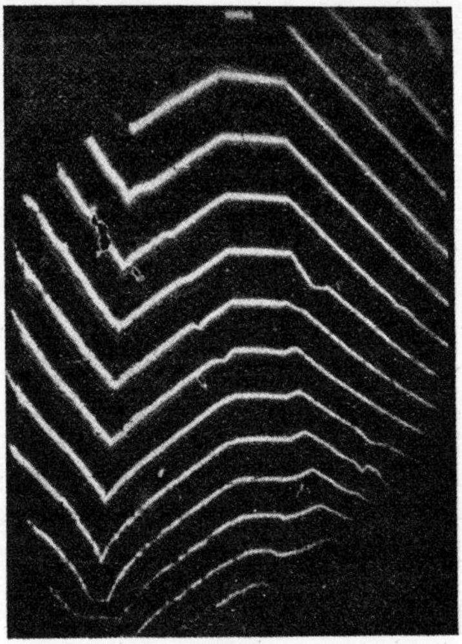

FIG. 27.

The fringe contour

Two typical contours, with low magnification, are shown in Figs. 26 and 27. Such pictures will be called *interferograms*. Although apparently different, they represent the same topography, as illustrated by Fig. 28, which represents the large ridge feature on Fig. 26. If this is intercepted by equidistant *horizontal* planes, as in the familiar contour maps of geography, the contour pattern obtained is that of Fig. 28*a*. If, on the

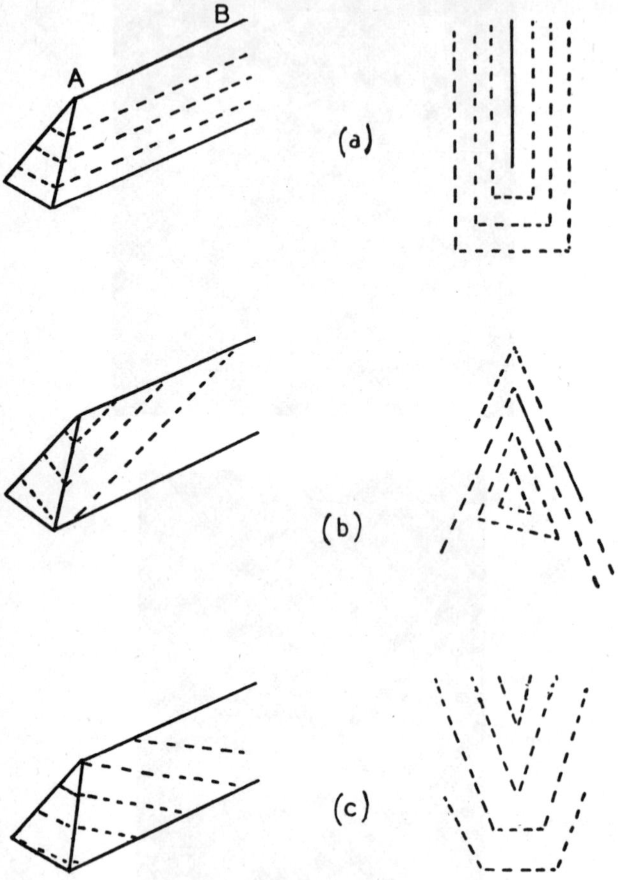

Fig. 28.

other hand, the feature is intercepted by equidistant planes making a positive or negative angle with the horizontal, then the respective contours are those shown in (*b*) and (*c*). This alteration of the plane of reference of the contours is obtained by tilting the flat relative to the crystal, and a set of such pictures can materially assist the interpretation. An improvement on this composite building up of features will be described later.

The first fact to decide is which are hill and which are valley features. This is settled by pressing lightly on one side of the crystal, bringing flat and crystal nearer together at that point. The fringes move towards the thicker part of the air wedge and thus ambiguity about hill and dale features is resolved. It is not always possible to apply this simple procedure and alternative methods are available and will be described later.

The next important feature is the determination of surface angles, and to achieve this a number of methods may be used. The previous example will be considered, the object being the determination of the angle between the large faces of the ridge of Fig. 28.

In the first method fringes are set as in Fig. 29 *a*, in which, by suitable adjustment of the reference flat, the fringes on one side of the ridge run perpendicular to the ridge edge, AB. By producing the fringe XY to cut the next order at Z, it is clear that the surface containing Z has dropped (or risen) through $\lambda/2$ in the length YZ. The angle between the two faces is $\lambda/2YZ$, using the accepted convention that the angle between crystal faces is the angle between the normals. Correcting for the magnification of the image on the photographic plate, it is $M(\lambda/2YZ)$, where M is the magnification. This method is suitable for angles of the order of one minute of arc upwards.

A second method is to set the fringes roughly symmetrically, as in Fig. 29 *b*, and to measure the distances XY and YZ. The angle between the faces is approximately

$$\frac{\lambda}{2}\left(\frac{1}{XY}+\frac{1}{YZ}\right)$$

for supposed unit magnification. If YZ and XY are approximately equal (and this is a matter of setting the flat), then an error in drawing ZX normal to AB is of little consequence, for the increase in length of YZ is almost exactly compensated by the decrease in XY.

A third and general method (due to W. L. Wilcock) requires

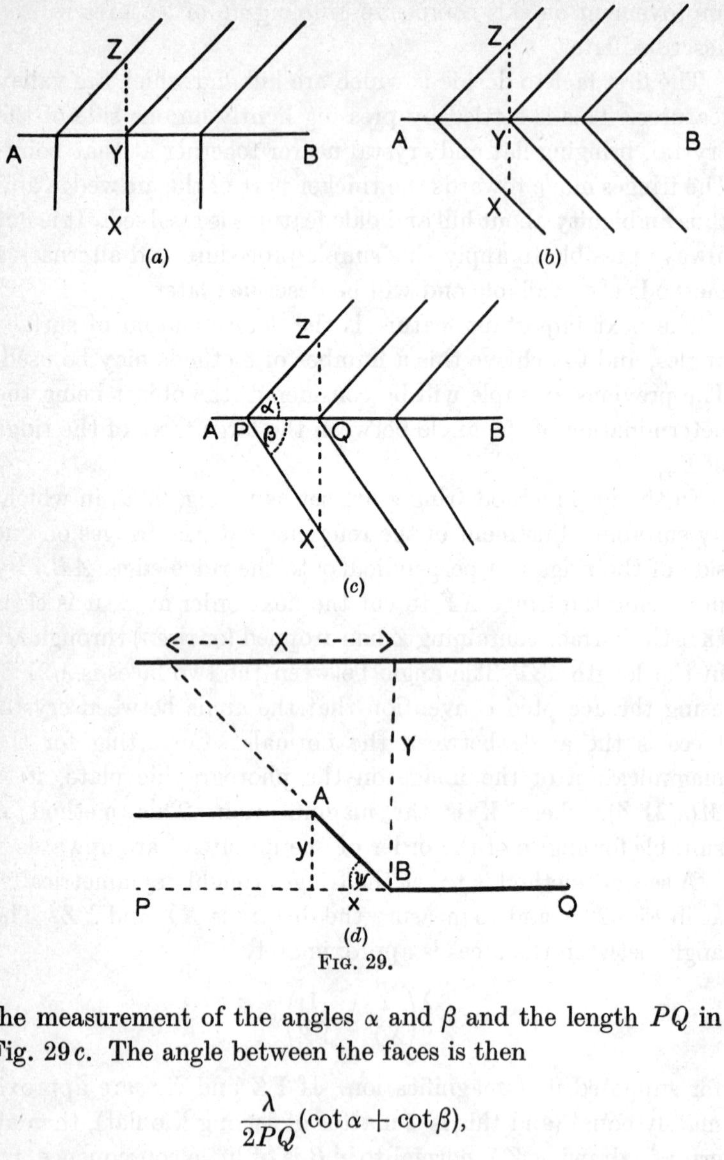

FIG. 29.

the measurement of the angles α and β and the length PQ in Fig. 29c. The angle between the faces is then

$$\frac{\lambda}{2PQ}(\cot\alpha + \cot\beta),$$

which is equivalent to the previous case.

For quite a small local kink, such as at B in Fig. 29d, it is seen that by measuring ψ, the angle between AB and PQ, and the distance Y between adjacent fringes (the dispersion) then the local surface slope is

$$\frac{\lambda}{2X} = \frac{\lambda}{2Y}\tan\psi.$$

With a larger feature the former methods apply.

The sensitivity of the method will be evident. If ψ is small the slope may be written $(\lambda/2)(\psi/Y)$, in which Y can be made 1 cm. (high dispersion fringes as in Fig. 26). Suppose that a ψ value of 1° is measurable, then the angle measurable is approximately $\lambda/100$, corresponding to only 0·12 second of arc. It is to be noted that the smallest angle measurable over a centimetre of face with a goniometer (Rayleigh limit) is λ, and *this also presupposes a strip broad enough to give a signal distinguishable from other signals.*

The technique leads to a *magnification of detail which is great in one direction only*, i.e. in the direction of height or depth. The whole contour of the crystal shown in Fig. 25 is encompassed within a depth of some four light-waves, i.e. 1/5000 cm. Observed angles between various faces on it vary from 0·50 to 9·00 minutes of arc. Variations exist on moving along the length of a ridge, and it is further quite clear from Figs. 26 and 27 that the sides of some of these ridges are curved, e.g. the radius of curvature of the main left-hand ridge feature is of the order of 60 metres.

It is of interest to draw attention to the sensitivity of this procedure for the determination of a small angle. It should be noted too that very small angles can be evaluated over quite small areas of the surface. Comparison with the Rayleigh limit is of interest. For a face 1 mm. wide the smallest angle that can be measured by simple reflection (Rayleigh limit) is 2 minutes of arc; for narrower faces the angle is proportionately greater. It will be noticed that 1 mm. corresponds to about 1/10th of the whole pattern registered in Fig. 27. No goniometer could possibly show up such fine detail, in particular the detection of such features as variation of angle *along* a ridge. The superiority of these measurements over those obtainable with the goniometer is of course due to the employment of multiple beams. The Rayleigh criterion for resolution applies to $\sin^2$ intensity distribution, and it is the employment of a much more advantageous intensity distribution which permits such a higher degree of resolution to be achieved.

It seems clear that these major features must be regarded as fine-scale vicinal faces.

It may be noted that plane faces are the exception rather than the rule. Despite the apparent considerable scale of the major contour features, when the enormous magnification is recalled it will be recognized that in fact the crystal face is fairly flat, since contours extend only over four light-waves.

Minor features

Comparison between the various fringe patterns reveals data concerning the minor markings on the surface. Interpretation shows that these markings are mainly of three broad types, namely (i) simple almost parallel striations, (ii) ankle bend marks of the form $\wedge$, which are a characteristic feature of the (100) face of quartz, and, (iii) features closely similar to (ii) but with the base closed to form an obtuse triangle, so, △.

(i) *Striations*

The striae are only well marked on the sides of the major ridge. Some are much fainter than others. Where the fringes pass over the more clearly marked striations a small kink appears. Each observable kink is a significant detail, correlating with a visual striation mark. From the direction of the kink (whether the acute angle is in the direction of increasing or diminishing order of interference) it can be decided whether the striation represents a rut or a small ridge. Measurement shows that the depths (heights) of striae are between 0·04 and 0·05 of an order, i.e. only some 100 A.U., which corresponds to 20 silica molecules.

(ii) *The* $\wedge$*-shaped markings*

Little information can be directly derived about these markings. A discontinuity in level takes place in their neighbourhood.

It can be concluded that in the near neighbourhood of the marks there are abrupt small changes in level, of the order of a small fraction of a wave.

(iii) *The triangular markings*

Data of some interest were obtained about these. Complete triangular fringe patterns are visible and represent small tetrahedral pyramids with their bases resting on the surrounding main surface. These tetrahedra are projections, not pits. Measurements made on one of these gave some 0·17 of an order as the approximate height of the pyramid, i.e. some 450 A.U. As the size of the pyramid base is a relatively large, easily seen triangular marking, it follows that the faces are inclined at very small angles to each other.

A significant point is that the small triangles closely resemble the major triangles. The face angles on the small pyramids are therefore close to those enclosed between the sides of the major feature; the pyramids can be regarded as a ridge meeting a triangular facet, or alternatively the major ridge feature can be regarded as a tetrahedral pyramid. It is reasonable to conclude that both features are identical but on different scales. It is probable that the small pyramidal projections represent the nuclei of growth of the (100) face, which have of course been arrested. The crystal growth of layers from such nuclei has been recently directly observed in many crystals. (Further structural details are shown in the reflection interferograms, Figs. 91 and 92.)

That the visual markings have been proved to have depths (heights) which are only a very small fraction of a wave-length is a matter of some interest optically. The fact that they are so easily seen requires explanation. It cannot be assumed that the markings appear only because of a differential reflectivity between the marked area and the surrounding surface. This follows as the various markings can still be clearly seen after silver has been deposited on the crystal face. The reflecting coefficient is then uniform, i.e. that of the silver. Proof of this lies in the fact that the fringes retain their sharpness whilst passing over the relatively large marked features. It can be concluded that a differential surface quality enables the markings to be seen. If it be assumed that the surfaces within the areas of the markings are ribbed or striated, either microscopically or even submicroscopically, then in a sense they will behave as crude gratings, whilst the surrounding lustrous areas act as plane mirrors. Two facts support this explanation. The angle of illumination required to render the markings clearly visible with a hand lens is somewhat critical and small changes in the angles make the markings appear to change over from bright to dark, a typical grating characteristic. Furthermore, the markings are more clearly seen when the crystal surface is viewed through a colour filter. It is only necessary to postulate the association of a high lustre with the true (100) face to account for the difference in character of the faces of the small tetrahedral pyramids, since the latter are vicinal faces.

It is possible that some of the observed markings can be seen because of the light diffracted from their edges, in accordance with Rayleigh's views on the visibility of discontinuities. This is a very probable explanation of the visibility of the ridge edges (junctions of vicinal faces) and of the striated markings. It is not likely to be the explanation of the visibility of the small tetrahedral pyramids since these extend over appreciable areas and it is quite clear visually that the whole of the area, and not only the boundaries, has a marked non-lustrous character. If the pyramids consist of stepped ridges, and this is probably the case, then in a sense the Rayleigh mechanism can be considered to operate from each sub-microscopic step. In this case the final effect is not very different from that of the pseudo-grating considered above.

It will be seen that the interference technique is likely to provide information concerning crystal growth in general if applied to this subject.

Mica cleavage

The previous section described the application of multiple-beam methods to the examination of a natural crystal face. In what follows it will be shown that when applied to cleavage faces of crystals much information is revealed. Mica will first be considered since it has a perfect cleavage, the surfaces of which possess a high natural polish. The procedure here described can be applied widely to many of the other crystals exhibiting perfect cleavage. A well-cleaved piece of good quality

FIG. 30.

mica has a smooth surface, but on it, cleavage 'lines' are clearly visible with a hand lens, frequently radiating out in an arbitrary direction from an area close to the point of insertion of the needle used to initiate cleavage.

A freshly cleaved mica surface is quite clean and no cleansing treatment other than ionic bombardment degassing is used when silvering. Typical interferograms from the surfaces of muscovite mica sheets are shown in Fig. 30 (magnification $\times 2$) and in Fig. 31 (magnification $\times 15$).

When thin sheets of silvered mica are mounted close to a silvered optical flat they are found to behave as highly sensitive diaphragms and are susceptible to small air shock-waves. The fringes are rarely quite still. This is particularly noticed in regions of high dispersion. (So sensitive a diaphragm, which indicates displacements of the order of $\lambda/100$, might have some useful applications.) The difficulty may be overcome by lightly pressing the mica against the flat, with a sheet of glass. Then the separation between the silvered surfaces is very small and

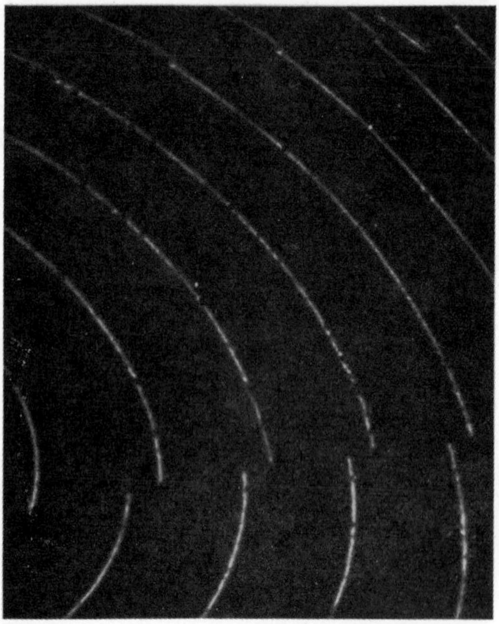

FIG. 31.

the fringe definition is excellent. This affords complete stability over long periods, but there is loss in control in adjusting the tilt of the mica surface relative to the flat. Sufficient control is provided, however, by light local pressure on the glass backing.

The stronger fringes in Fig. 30 belong to the green mercury line, the weaker doublet being the yellow lines. To assist identification, photographs are often taken without filters. Despite the obvious complexity of the patterns, three characteristic salient features are evident, and such features appeared on all the mica samples examined.

(i) The surface contains hills and dales (in some cases quite small pits). Different samples vary only in degree, but not in character. Fig. 30 shows, for example, numerous hills, some 1/200th mm. in height, whereas the sample of Fig. 31 is much more nearly plane.

(ii) A striking characteristic is the existence of sharp clear-cut cleavage lines representing discontinuities in level. These vary considerably in length and are often fairly straight. In some samples they run in one general direction, in others they intersect, at times with large angles up to almost 90°.

(iii) There exist large areas between the cleavage lines which are of great optical uniformity. One must distinguish between (*a*) distortion into hills and dales, and (*b*) local irregularities. The smooth run of the fringes between the cleavage lines shows that, although the surface may be badly distorted locally into a sharp curve, it remains smooth and uniform. Of interest is the manner in which some cleavage lines pass straight over hill-and-dale features.

Owing to the manner of mounting it is not easy to determine which are hills and which are valleys. This can be achieved, however, by two methods. One method employing white-light fringes is described later. The second is that developed by J. Brossell [24] and is based on the following.

The interference within the air gap obeys the relation $n\lambda = 2t\cos\phi$, in which n is the order of interference with a film of thickness t, λ the wave-length, and ϕ the angle of light incidence. For a point source at the focus of a lens, giving a parallel beam at normal incidence, sharp fringes are formed as shown in Fig. 32*a*. Suppose now an extended source replaces the point source, then values of ϕ greater than zero (the normal incidence value) now exist in the beam. The fringes thus broaden, but on one side only, namely in the direction of increasing t. This occurs because $\cos\phi < 1$ for all beams other than normal, so that for these a given order n requires a larger value of t, which serves to reveal the directions of hills and dales. The effect of using a considerably extended source with the same mica surface as used for Fig. 32*a* is shown in Fig. 32*b*.

The hill-and-dale features reveal much of interest when a systematic examination is made of different varieties of mica, but this aspect will not be discussed here.

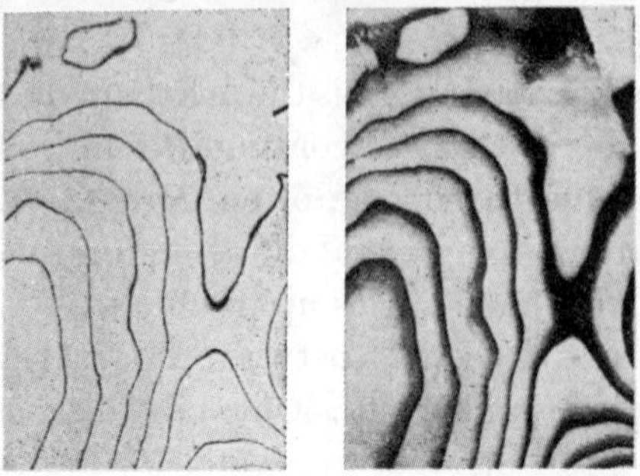

(a) (b)

FIG. 32.

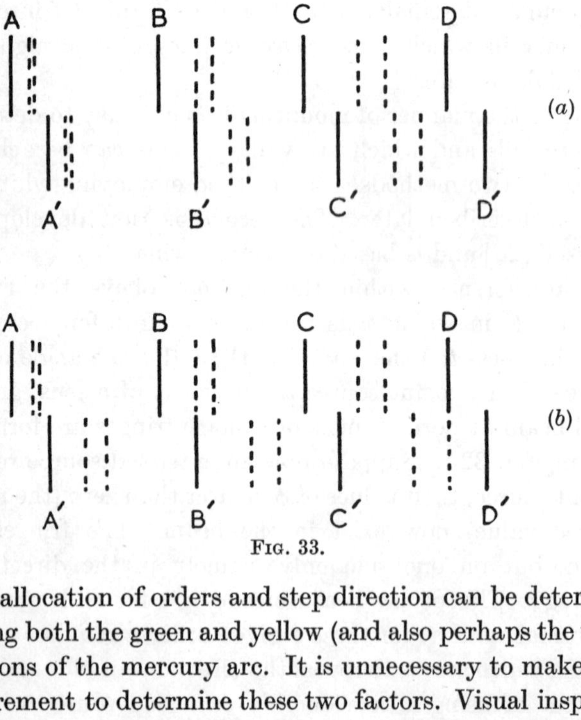

FIG. 33.

The allocation of orders and step direction can be determined by using both the green and yellow (and also perhaps the violet) radiations of the mercury arc. It is unnecessary to make exact measurement to determine these two factors. Visual inspection of the pattern often suffices. This is made clear by Fig. 33, in which the strong line is the green and the dotted ones the yellow. In Fig. 33 *a* the lower system $A'B'C'...$ is displaced to the *right* from the system $ABC...$ by the amount $AA' = BB' = CC'...$.

In Fig. 33 *b* the grouping A' is the same as that of B, B' the same as that of C, etc.: hence the lower system is displaced to the *left* by the amount $A'B = B'C = C'D...$.

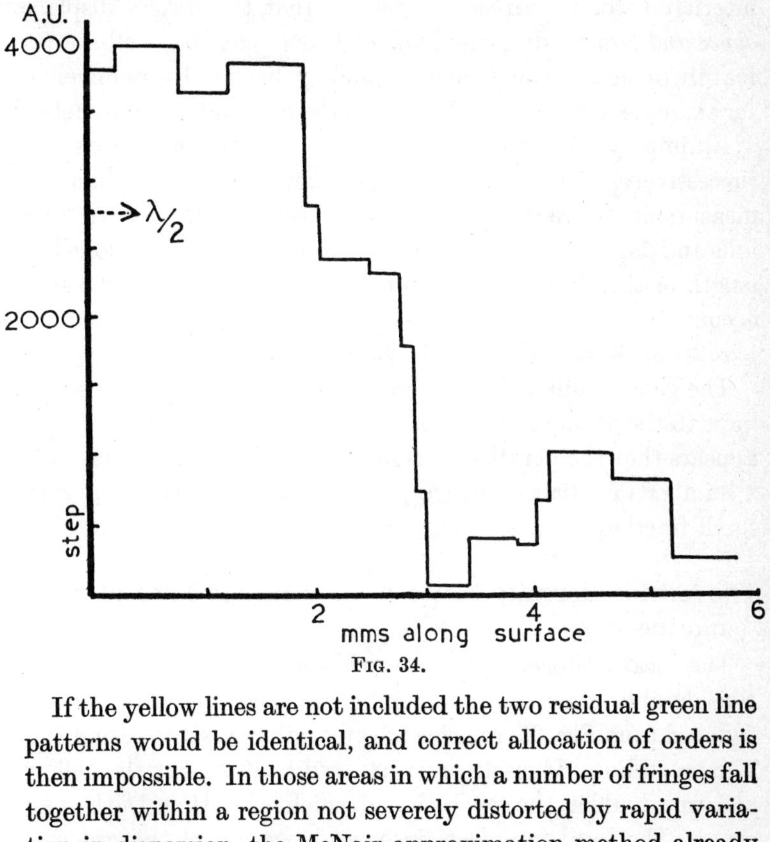

FIG. 34.

If the yellow lines are not included the two residual green line patterns would be identical, and correct allocation of orders is then impossible. In those areas in which a number of fringes fall together within a region not severely distorted by rapid variation in dispersion, the McNair approximation method already described permits a precision determination of the step to be made.

The contour of a traverse across 6 mm. of a typical cleaved sheet of muscovite is shown in Fig. 34. In the best of cases it is possible to measure the step to within an accuracy of 3 A.U. This requires the measurement of the fractional order displacement to within 0·001. This can be done, for although the fringe width is about 0·02 of an order, yet owing to the method of setting, a displacement measurement can be made with an error of no more than 1/20th of the fringe width.

It is important to note that both resolving power and precision considerably exceed that in the corresponding Fabry-Perot interferometer by virtue of the fact that the fringes displace *above and below* a dividing line. This is of course a well-known feature in measurement, such a fiduciary displaced system being, for example, used in a similar way in the Rayleigh refractometer.

An important feature concerns the value of a step measured successively from a run of fringes along a cleavage line. If measurements are made in those regions not badly distorted by hills and dales, it is found that the fringe displacement along the length of the cleavage line is uniform, indicating that the step is constant, so that *the two faces separated by the cleavage line are parallel* or, if curved, have the same curvature.

The general hill-and-dale contour of the whole surface sets a limit to the accuracy with which this can be determined, but it appears that the parallelism is fairly exact locally, at least over a number of adjacent fringes, and certainly to within a very small fraction of an order.

The local uniformity between the cleavage lines of mica and the steps

The local uniformity between cleavage lines is a feature of considerable crystallographic interest. The following can be deduced from Fig. 35, relating to three different samples of mica. The green line fringes are over-exposed to bring out the yellow mercury doublet, the separation of which is less than 1/7th of an order. The local 'wriggles' in each fringe are of the order of perhaps 1/15th of the doublet separation indicating local fluctuations of the order of about 1/100th of a fringe. This amounts to 1/200th of a wave and is clearly the limit set by the optical flat. Thus it can be concluded that, disregarding general superposed curvature, the mica surface must be uniform to better than 30 A.U.

Sir W. L. Bragg (private communication) has pointed out that these cleavage lines and steps are almost certainly to be identified with the invisible steps inferred to exist on some mica surfaces from experiments made by Friedel on the growth of ammonium iodide crystals upon such surfaces (see Bragg,

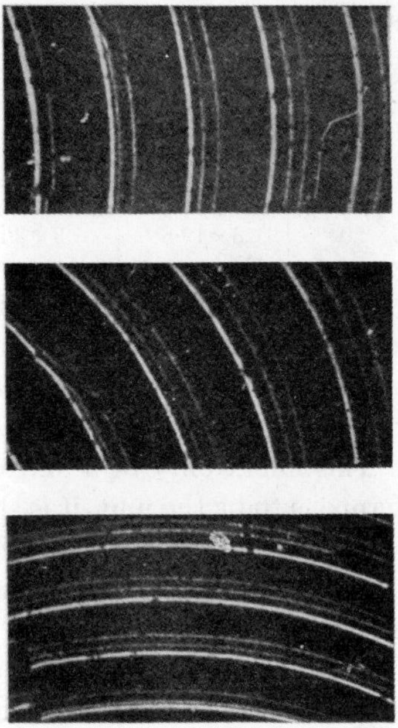

FIG. 35.

Atomic Structure of Minerals, 1937, p. 210). On some samples the crystals orient themselves oppositely along invisible dividing lines, and it has been concluded that over the area of identical orientation the mica surface has cleaved true to within a single molecular plane. According to Bragg, as the mica cleaves along the plane in which the K atoms are situated, and as the height of the molecule (strictly the c spacing) is 20 A.U., it follows that all the cleavage steps should be integral multiples of 10 A.U., since the K atoms are centrally situated.

The interference data can now be compared with these conclusions. The smallest steps recorded from mica interferograms are 20 A.U. in height, i.e. exactly the *c* spacing (effectively one 'molecule').

The following typical set of measurements show five steps in which the precision of measurement is high, since they occur in areas nearly plane. The fractional order displacement is known to 0·001, which corresponds to an error of the very small value of 3 A.U. It is a striking fact that *these steps are exact integral multiples of* 20 A.U. (within the error). In Ångstrom units they are

Steps in A.U.

41	*100*	*158*	*180*	*341*
$2\times 20{\cdot}5$	5×20	$8\times 19{\cdot}8$	9×20	17×20

Furthermore, these steps all being *small* multiples of 20 A.U., there is no question about the certainty of the integral ratios. *It is clear that the steps are simple multiples of whole 'molecules'.* There is amongst these five, no evidence of 10 A.U. being the fundamental unit. This might only be a matter of chance, for clearly if 10 A.U. happens to be the unit, it is simply a question of whether amongst five random steps all will have an even number of units. Other evidence seems to incline to favour 20 A.U. as the unit.

It should be noted that this in no way contradicts the ammonium iodide crystal growth experiments. The successive double layers of mica are related by a glide plane of symmetry parallel to (010). The orientation on the surface for any step may have an inclination on either side of this symmetry plane. On adjoining steps the orientations may be the same, and then the crystals of ammonium iodide experiments would in fact not reveal a step. Three illustrations of crystal growth are given by Bragg. One, lepidolite, shows a clear step, hence this is a half-integral multiple of 20 A.U. But another sample shows no 'invisible' step. It does not follow that the surface is uniform, it may in fact have on it a number of steps, each an integral multiple of 20 A.U. The mica samples used here would give ammonium iodide patterns similar to this second sample. The third crystal growth illustration (phlogopite) shows orientations of either kind equally numerous, from which it might possibly be concluded that here we have a typical area in which there are many small-length cleavage steps of narrow width between steps. Such characteristic areas have appeared on some of the samples examined interferometrically.

Not only is there agreement between the general conclusions as regards steps, obtained by interferometry and by crystal growth experiments, the same is true of surface uniformity. The existence of uniformity between the invisible cleavage lines given by the superposed crystal growth orientation is not quite a safe inference, for, as already pointed out, there exist steps which are integral multiples of 20 A.U., and from the point of view of crystal growth such steps would have no effect on orientation and thus be missed, giving a false picture of uniformity.

It was concluded from the smoothness of the doublet yellow fringes that, between cleavage lines, the surface is, however, actually uniform to at least 30 A.U. Thus the uniformity can be objectively demonstrated with fringes, although only deduced from the crystal orientation (though the orientation observations are subject to the doubt already mentioned). It seems that the conclusion drawn from the crystal growth, namely that between the cleavage lines the surface runs true to a molecular plane, is substantiated.

Decisive evidence on this point is given in a later section, where it is shown with certainty that some mica samples cleave true to a single molecular plane over areas exceeding 20 sq. cm.

The silver contour

A feature of fundamental importance to the whole technique is illustrated both by the smoothness of the fringes and the uniformity of the cleavage step value observed along the length of a line (see, for example, Fig. 31). Measurement shows that, along the length of a cleavage line, the value of the step is constant to within limits of error. Crystallographically this was to be expected. Yet its observation yields information on a critical point. It follows from the measured uniformity of step that the silver must have *exactly* contoured the surface, probably to within the crystal lattice distance of silver. What is meant by this statement is illustrated in Fig. 36.

Suppose $ABCD$ is the mica surface and that the silver deposits itself as $A'B'C'D'$. One can, for convenience, imagine specular reflection taking place from some equivalent surface $A''B''C''D''$.

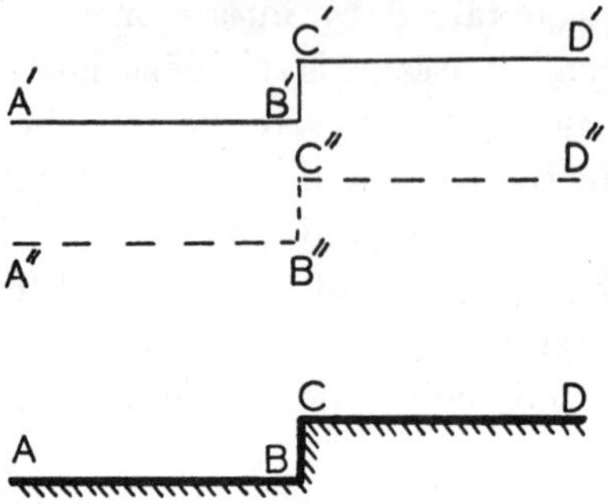

FIG. 36.

(Strictly speaking, reflections take place within the whole depth of the film, but it is unreasonable to push the application of this model to its limit.)

It appears certain that the difference in level between $A''B''$ and $C''D''$ (also that between $A'B'$ and $C'D'$) is identical with that between AB and CD. In this sense the silver exactly contours the surface. It is not postulated that the deposition locally at the edge $B'C'$ (or $B''C''$) has the same shape as BC—indeed it is highly probable that it has not. For this reason a limit is set to the *linear* magnification that can be used in viewing fringes. Experiments have been carried out on crystal features of diamond which indicate that $\times 400$ can be tolerated with no evidence of serious disturbance from silver variation. This would imply that any irregularity round $B''C''$ has a relatively small lateral extension.

Further evidence will be adduced later proving that the silver contouring is adequately perfect for the technique.

The cleavage of selenite

Selenite crystals (gypsum) have a perfect cleavage so closely resembling that of mica that in the early history of crystallography the two materials were frequently confused. Yet when the cleavage surfaces are examined by multiple-beam interferometry, the differences are strikingly evident and reveal in a clear manner the potency of the technique.

The specimens of selenite available gave fairly good specular

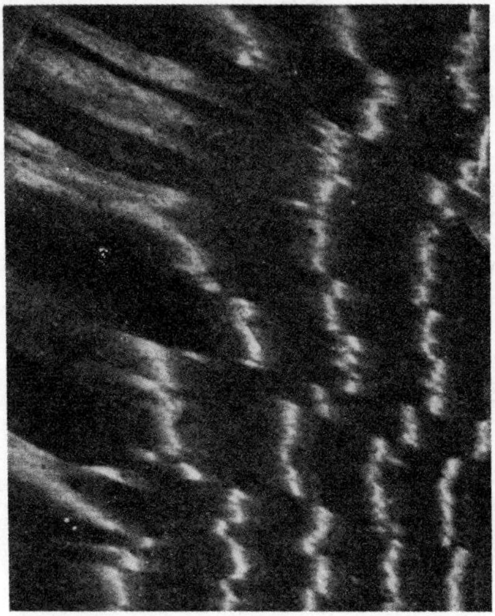

FIG. 37.

reflection (before silvering) over some 2 sq. cm. Typical contours, representing about 1 sq. cm., are shown in Figs. 37 and 38, and contrast strikingly with the mica ones. The main distinguishing features of the patterns are:

(i) A large number of roughly parallel cleavage lines. The large angles between some cleavage lines exhibited by some micas do not appear, although in isolated cases angles up to 15° appear.

(ii) The characteristic hillocks and valleys shown by all mica specimens are absent. Instead the selenite shows a cylindrical curvature. The curvature is relatively large, the radius being some 4·6 metres, for one particular sample. The cleavage lines are in the same direction as the curvature.

(iii) The separate fringes are very ragged and show multiple kinks. Between the straight cleavage lines the fringes are nowhere smooth, as was the case with mica, but are disjointed and irregular, although retaining their individualities as separate, reasonably sharp, fringes (Fig. 37 particularly).

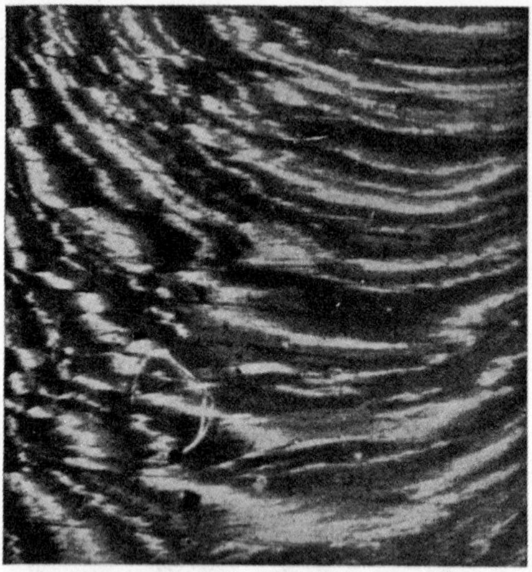

FIG. 38.

The detail of the separate fringe structure is shown in Fig. 38, where the dispersion has been increased by bringing the crystal surface nearer to parallelism with the optical flat.

The cleavage-line discontinuities in selenite

The ragged nature of the fringes reduces the precision of measurement compared with that for mica, though it is still considerable. The contour of a typical traverse across 7 mm. of the crystal, at right angles to the cleavage lines, is shown in Fig. 39. The numbers of 'molecules' in the steps are approximate whole numbers, for the accuracy is insufficient to give an exact figure.

There exists a close superficial resemblance with the typical mica contours. A very large number of steps smaller than those recorded above were observed, but they are not illustrated in Fig. 39 since they extended for only short lengths and tended to become merged into the general ragged nature of the fringes. It will be noted that the X-ray dimension of the selenite cell in the direction perpendicular to cleavage is 15 A.U., which compares with the 20 A.U. of mica.

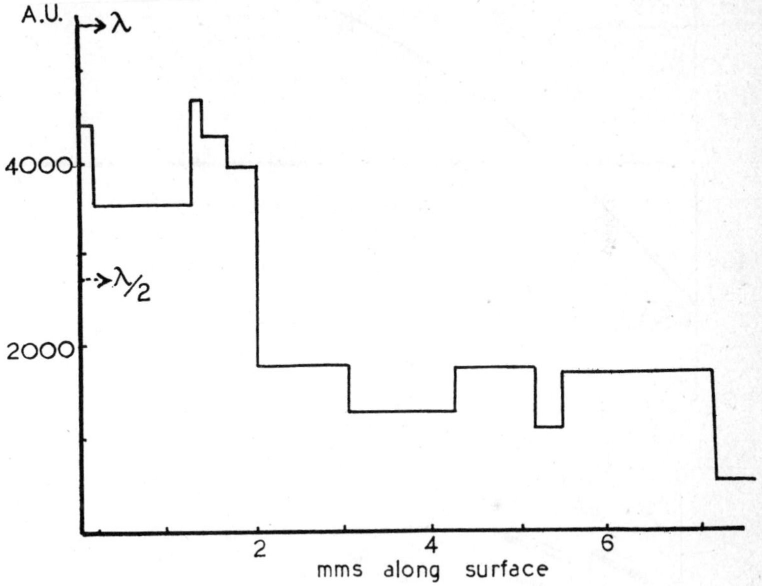

FIG. 39.

Variation of step along a selenite cleavage line

A marked difference between mica and selenite is that the value of the step, in many cleavage lines, changes regularly on moving along a line. This means that the two surfaces separated by the cleavage line are not parallel, having a relative slope along the direction of the line. A particularly marked case is shown by the data given in Fig. 40, which is the plot of the value of the cleavage step along a 3 mm. length of a cleavage line. This shows that there is a small relative curvature between the two faces superposed on a general slope. Taking one surface as the horizontal plane, the other starts off below this, crosses over, and a downward step becomes an upward step.

The case illustrated in Fig. 40 is an extreme example, for in the majority of the cleavage steps measured any relative

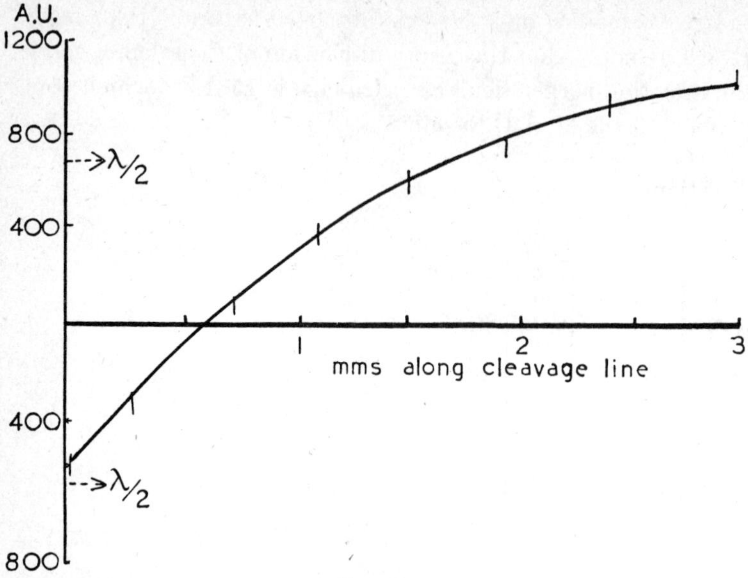

FIG. 40.

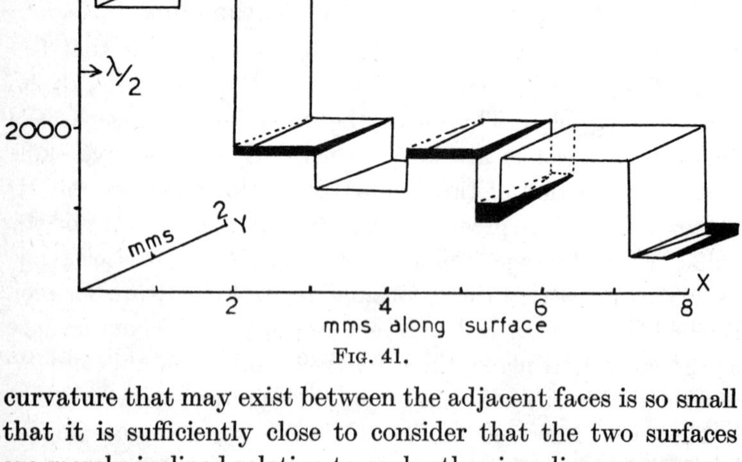

FIG. 41.

curvature that may exist between the adjacent faces is so small that it is sufficiently close to consider that the two surfaces are merely inclined relative to each other in a linear manner.

The character of the surface is illustrated by an approximately perspective diagram, Fig. 41, which represents the contour with the third dimension added. The axes OX, OY, OZ are mutually perpendicular, the scales along the two horizontal directions are the same, the magnification being great only in the vertical direction. The deviations of the surfaces from the horizontal plane are shown by the blackened wedge sections. The thickness of the slab along the OY direction is 1 mm.

The angles between the inclined faces are very small, and it is only the high magnification in the Z direction which enables them to be measured. It may be noted that an angle of 0·011 minute of arc for one of the strips is given by a crystal strip only 0·8 mm. long and 0·4 mm. wide.

The state of the surface between the selenite cleavage lines

The peculiar state of the surface between the long cleavage lines is revealed by the higher dispersion regions, especially near to the centre of the cylindrical fringe pattern. The figures show why with lower dispersion the fringes have so ragged an appearance. It can be seen that the areas between the major cleavage lines consist of a large number of small facets, each of which is an elongated strip. The long axis of the strip is in the direction of the major cleavage lines. The strips vary in length from quite small strips to lengths up to 1 mm. The widths also vary widely, and although most are of the order of only 1/50 mm. wide, some attain a width of the order of 1/10 mm. These strips do not lie in the same plane but are stepped in a haphazard way, positively and negatively, the steps being only small fractions of an order in most cases. It is clear that the strips differ in height by only a few molecules. It will be shown later that many steps are only one molecule high.

There is no evidence of any regularity in this distribution of small strips, or in their heights. Occasionally an individual narrow strip will project out of a fringe by as much as 1/10th of an order, indicating that there is a difference in height by some 300 A.U., i.e. some 20 'molecules'. In general the differences are considerably less.

The individual small elongated strips practically all lie with their major axis in the general direction of the long cleavage lines, but in some cases there are appreciable abrupt angles between adjacent small strips. The general direction of the large majority proves that the strips must be regarded as subsidiary cleavages superposed upon the main areas defined by the major cleavage lines. The strip edges are sharp, indicating true discontinuity.

Selenite is a hydrate, and the crystal had been exposed to a vacuum for some 30 minutes during the silver-deposition procedure. It must be considered whether some slight surface dehydration effect might account for the ragged surface. The crystal never warms up above room temperature during the silvering procedure, so that the dehydration, if any, might be expected to be minute. On the other hand, the facets are only a few molecules deep, often only one molecule deep. It seems certain, however, that dehydration has nothing to do with the effect. The fact that the individual strips show such sharp cleavage edges, and are in the same general direction as the major cleavage lines, disposes of this view. The subsidiary strips must certainly arise during cleavage, indicating a true secondary structure.

Thus the cleavage mechanism differs from that of mica in that there are two distinct factors: (*a*) the major strong cleavage strips extending over distances of the order of centimetres; and (*b*) the secondary weak cleavage patches which are in most cases only a fraction of a millimetre in length. In mica the cleavage takes place at a weak bond between potassium (or sodium) and oxygen, whereas in selenite the weak bond is between water and oxygen. This difference may be connected with the different cleavages. On the other hand, whether or not the subsidiary patches in selenite are related to a crystal 'mosaic' remains yet to be determined.

The contrast between the mica and selenite cleavages is further accentuated by the common occurrence of a relative slope between the two faces on either side of a cleavage line in the case of selenite. It is also fairly certain that in selenite the change in level as one moves along a cleavage line is not continuously smooth but rather stepwise and ragged, because the subsidiary patches are so much shorter than the main cleavage lines and they lie at variable heights.

It is clear now that the three-dimensional diagram in Fig. 41 is only a simplified schematic representation of the true state of the surface.

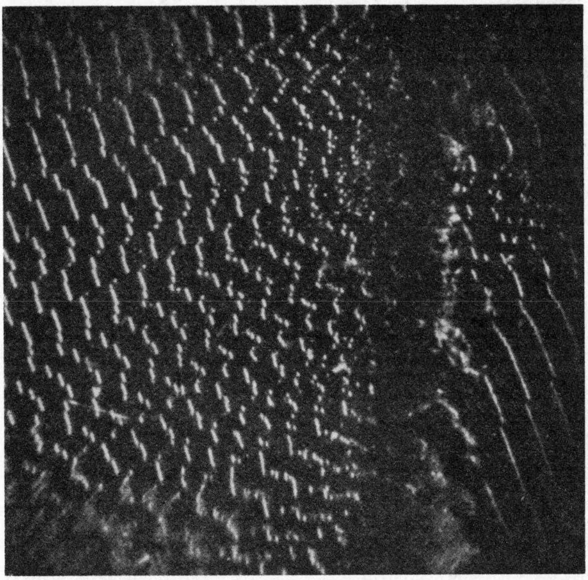

FIG. 42.

The cleavage of calcite

Calcite cleaves perfectly in three directions to form the well-known rhombs, and an examination of a calcite cleavage face has revealed a feature of crystallographic interest. The Fizeau fringes for the face in question are shown in Fig. 42 and are characteristic. The calcite exhibits a series of fairly parallel cleavage lines between which are smooth fringes recalling the smooth continuity of the mica fringes. It appears that between the major lines the calcite *might also be expected to cleave true to a molecule.* Since the lattice spacing is only 6 A.U., this dimension cannot be subject to numerical confirmation. Some perfectly uniform areas extending over 2×2 mm. have been observed. Fig. 43 is a print with fringes running nearly parallel to the cleavage lines.

It is probable that the cleavage pattern is evidence of some form of mosaic block pattern within the body of the crystal,

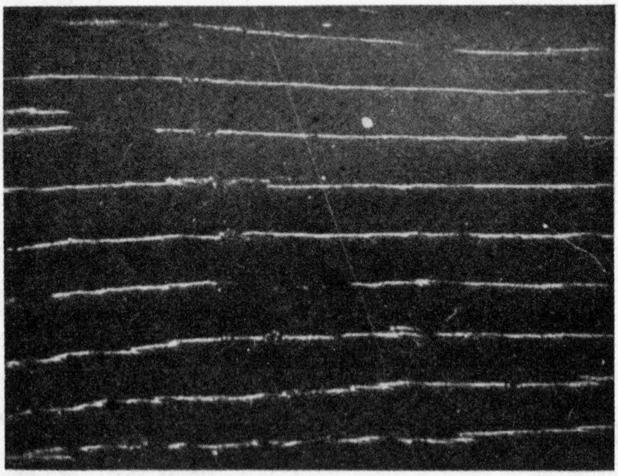

FIG. 43.

somewhat in the nature of the lineage formation postulated by Buerger. Thus a highly uniform cleavage area over some considerable extension is probable evidence for the existence of a local block of a perfect single crystal. By this means, therefore, it should be possible to select a high-quality crystal for use in an X-ray spectrometer, since the calcite crystal is used for this purpose.

The majority of the cleavage lines resemble those in mica in that the value of the cleavage step remains constant along the length of the line; but in rare instances this is departed from. In a few particularly noteworthy cases a curious effect has been observed. Application of light pressure to the crystal surface (by pressing on the optical flat) alters the heights of these particularly sensitive cleavage strips. In one of the narrow strips (strip dimensions 2 mm. × 0·1 mm.) bounded by two cleavage lines, pressure causes the whole strip to move as if rotating about one end, the strip dipping down into, or rising up out of, the body of the crystal by a very small amount. Each successive application of pressure leads to an unpredictable movement of the strip. A typical movement can lead to a change in level of some 100 A.U. The strip appears to be *gliding* through this very small distance, tilting through an angle of 1″ of arc.

Cleavage proceeds along a plane until the boundary of that particular single crystal is reached. At this boundary the cleavage jumps and a step results. The gliding of individual blocks mentioned above supports this view. It is highly probable that the perfect rhombohedral cleavages of calcite permit this gliding to take place. The observed shapes of the cleavage strips vary considerably in different crystals, being broad rectangles in some and narrow strips in others (e.g. 3 mm. × 0·03 mm.).

A tentative mechanism for the cleavage can be proposed on the basis of these observations. Within a single crystal, cleavage once begun carries on over a true molecular plane until reaching the boundary. At this discontinuity the cleavage jumps and a cleavage step forms. If this view is correct, the extensive cleavage areas in some micas are evidence of the formation of large blocks of perfect single crystals.

It will be seen from the discussion that the multiple-beam interference method has already thrown some light on the mechanism of cleavage, and if intensively applied should give information about the whole problem of surface structure and mechanical strength.

CHAPTER VI

CROSSED FIZEAU FRINGES

DIAMONDS

Microscope technique

THE multiple-beam fringes can be examined with a microscope provided the phase conditions previously discussed are not violated. It is clear that to view high-magnification Fizeau fringes the wedge angle must be increased if several fringes are to cover the field of view, which is usually essential.

If fringes are to be 1 cm. apart upon a microphotograph at $\times 100$, the distance between fringes on the silvered surface is 0·01 cm., and this necessitates as small a value of t as can be attained if the phase condition is to be satisfied. The angle θ between the surfaces is $\lambda X/2$ where X is here 100, which makes $\theta = 0{\cdot}15^\circ$ for the green mercury line.

Now the nth beam makes an angle $\xi = 2n\theta$ with the first beam and for $n = 60$ this gives $\xi = 18^\circ$. The light entering the microscope is unsymmetrically distributed, being all on one side of the direct beam; thus to a rough approximation a cone of semi-angle 18° is to be collected. This requires a numerical aperture of the order 0·45.

A good average half-inch objective has a numerical aperture 0·40, which is perhaps just on the limit for collecting most of the effective beams. Thus it can be concluded that for magnifications below $\times 100$, a $\frac{1}{2}$-in. NA $= 0{\cdot}40$ objective (with $\times 8$ eyepiece) suffices to collect most of the essential beams. For higher magnifications a bigger numerical aperture lens must be used, otherwise fringe definition will suffer. This is not a question of microscope resolution in the ordinary sense, although it is related to it, for clearly if fringes are 1/10 mm. apart the microscope is being expected to resolve 1/500 mm. if a fringe width of 1/50th order is to be sharp.

Fortunately the reduction of gap required for the phase condition permits the employment of an intense high-pressure mercury-arc source and as a result high light intensities are available permitting good exposures to be obtained in relatively short times.

Fizeau fringes with a diamond

The fringes obtained with diamond crystals illustrate the general application of the microscope technique. The diamonds

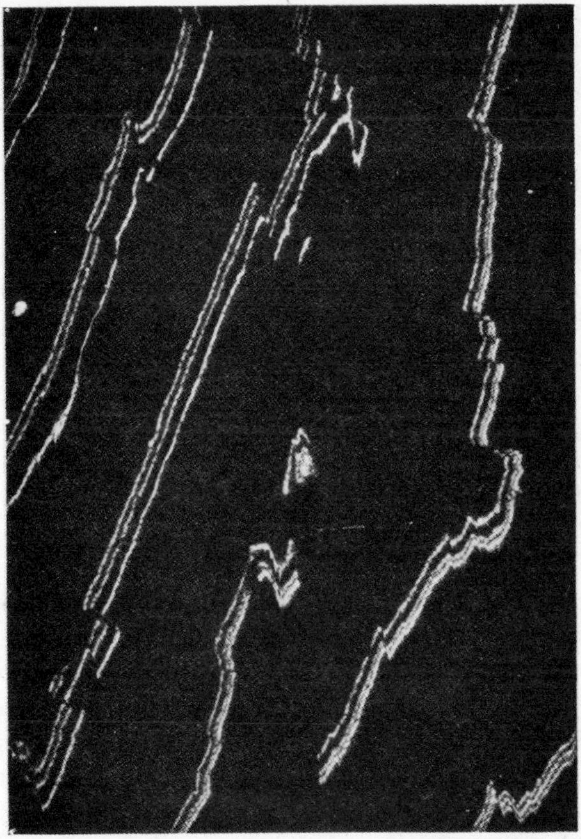

FIG. 44.

selected were 'portrait' stones, i.e. those in which a pair of opposite, parallel octahedron faces is abnormally developed, leading to a crystal in the form of a plane parallel plate. The diamond was cleaned, silvered, and mounted against a silvered flat, which could be brought as close as desired to the crystal surface, and with three light springs any required tilt between the surfaces could be obtained. The separation between the slightly inclined surfaces was reduced to a few light-waves.

The Fizeau fringes shown in Fig. 44 were photographed by means of a Leica camera attachment to the microscope, and green and yellow fringes are shown. The area covered was about 2 sq. mm.

Attention is drawn to the extreme sharpness of the fringes, and it will be realized that irregularities and steps of the order of a few Ångströms only can be measured with confidence.

However, the very sharpness of the fringes constitutes a defect in that only a small fraction of the total area is scanned by fringes on any single interferogram. To obtain a reasonably complete topographical survey it would be necessary to take a considerable number of such pictures, and from them (with a good deal of labour) the surface structure could be built up. This method was indeed used in examining the quartz crystal formerly described.

The *crossed fringe technique*, to be described below, overcomes this difficulty in a simple manner, reduces labour considerably, and adds a great deal to the power of the method.

High-dispersion Fizeau fringes

Suppose the wedge angle is reduced until the two surfaces are as near parallel as possible. Residual angles remain, due to the inherent topographical features. On removal of the colour filters a high-dispersion complex pattern, as in Fig. 45, results, and this exhibits a complete picture of the crystal surface, revealing with high contrast a wealth of structural detail not observable by ordinary micrographic methods. It is emphasized that Fig. 45 is an interference pattern of high sensitivity, as shown by the following considerations.

Fizeau fringes between parallel plates lead to a uniform tint of intensity determined by the separation of the plates in accordance with the Airy formula, maxima occurring for $n\lambda = 2t$, at which n is integral. A change in separation produces a change in intensity. Consideration of the fringe shape shows that the maximum sensitivity occurs if t is selected such that approximately half the peak intensity is in the field of view.

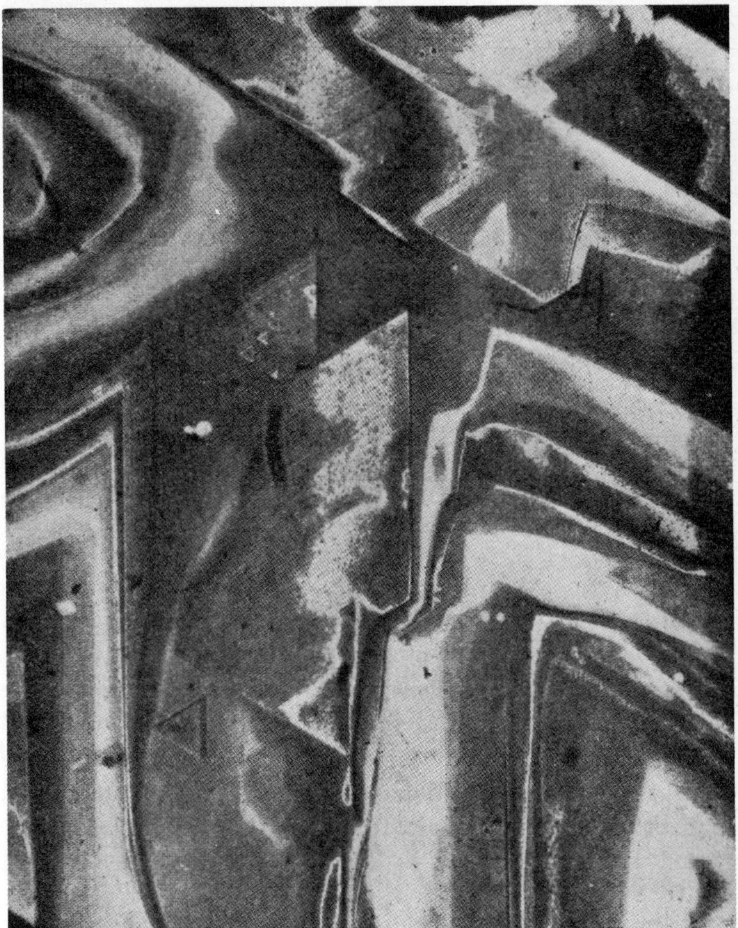

FIG. 45.

At this value $$I = \tfrac{1}{2}I_{\text{max}},$$

whence $$\tfrac{1}{2}I_{\text{max}} = \frac{I_{\text{max}}}{1+F\sin^2\frac{1}{2}\delta},$$

where $\delta = 4\pi t/\lambda$. Thus

$$\sin^2\tfrac{1}{2}\delta = 1/F.$$

Since F is large, $\sin\delta$ can be replaced by δ, giving

$$\delta = \frac{2}{F^{\frac{1}{2}}}.$$

Suppose now that a 10 per cent. change in intensity is produced by a change of t to t', corresponding to a new value δ' (where $\delta' = 4\pi t'/\lambda$):

$$0{\cdot}45\, I_{\max} = \frac{I_{\max}}{1+F\sin^2\frac{1}{2}\delta'},$$

giving
$$\delta' = \frac{2\sqrt{11}}{3F^{\frac{1}{2}}};$$

hence
$$t'-t = \left(\frac{2\sqrt{11}}{3}-2\right)\frac{\lambda}{4\pi F^{\frac{1}{2}}},$$

which equals $\dfrac{\lambda}{60F^{\frac{1}{2}}}$ very closely.

For $R = 0{\cdot}94$ this equals $\lambda/1{,}940$ since $F = 1{,}044$, for which, to a close approximation,

$$t'-t = 3\ \text{A.U.}$$

Since a 10 per cent. change in intensity is readily recognizable, the method is clearly one of extreme delicacy.

Yet the interferogram as it stands cannot be interpreted, for there is no indication as to any possible overlap of orders and in any case for numerical computation precise microphotometric density determinations would be required. The solution to this difficulty is given by crossed fringes, as described below.

Crossed fringes

If the ordinary sharp and the high-dispersion fringes are crossed, i.e. photographed in succession on top of one another on the same plate, then the result is as shown in Fig. 46.

The power of such a technique is strikingly obvious, for the various twists and turns of the sharp, narrow, precision high-definition fringes of Fig. 44 receive immediate interpretation and conversely the broad features of Fig. 45 can now be subjected to precise numerical evaluation.

A still more striking combination is shown by the *triple* crossed fringe system illustrated in Fig. 47. Here the high-dispersion fringes have superposed upon them two independent sets of narrow wedge fringes, placed roughly at right angles to each

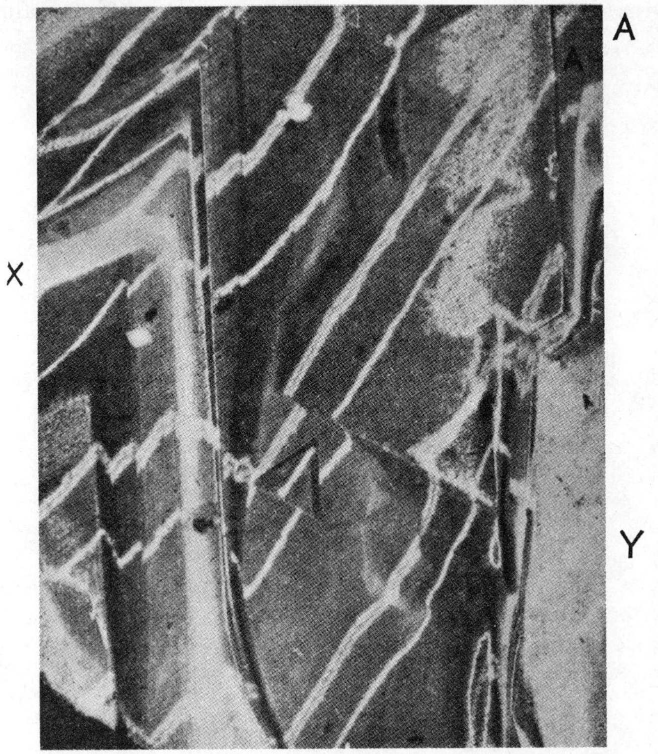

FIG. 46.

other, such an arrangement being obtained by successive suitable adjustment of wedge angles. This compound picture is so complete that a large number of important and interesting features can be numerically evaluated from the one photograph. The gain in ease of interpretation and the saving in labour is considerable.

It would be out of place here to go into the interpretation of the mass of detail shown by this picture, a report of which has been given elsewhere [26], but some of the main features will be discussed, illustrating, as they do, general principles.

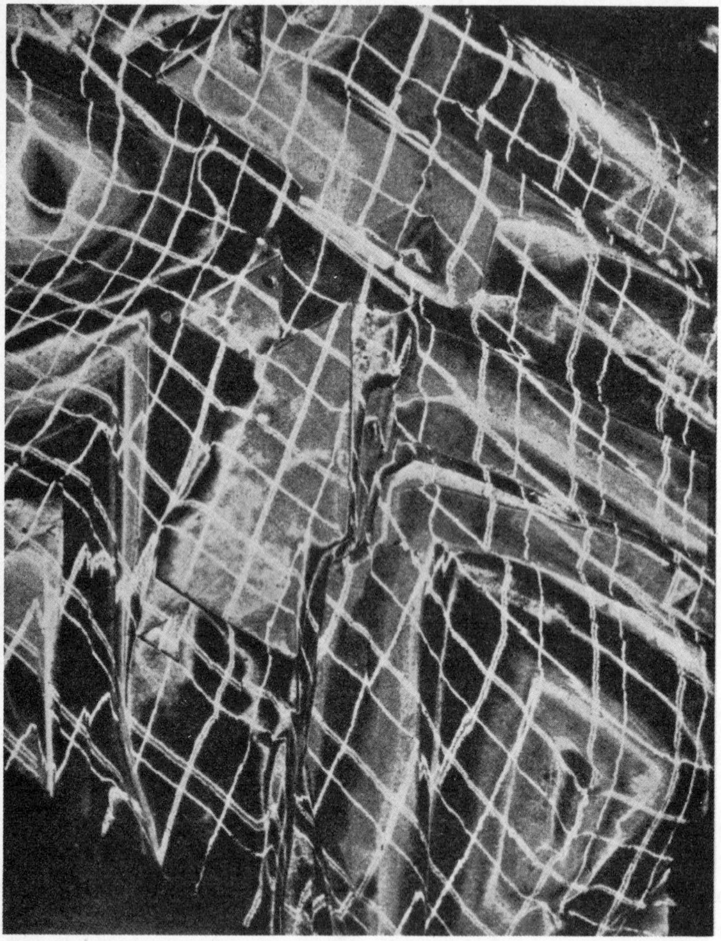

FIG. 47.

Triangular pits

There has existed some controversy for fifty years as to the origin of the shallow-pit triangular markings frequently seen on octahedral diamond faces. Many such markings are visible in

Fig. 45. One in particular will be discussed and is shown enlarged in Fig. 46.

It will be seen that the fringe beginning at A is straight until reaching the ridge marked XY. The application to one corner of gentle pressure shows that a displacement to the right means an elevation on the surface and identifies XY as a ridge. However, the fringe through the large triangle is a linear continuation of A. Thus the base of the triangle is at the same level (to within molecular limits) as the extensive outer area above XY. It is clearly entirely unreasonable to expect etching to go down exactly to this outer level. The triangle must arise from *growth* and is due to the incompletion of growth sheets in the region below XY. If it is postulated that growth takes place in three directions, inclined at 60° to each other, the arresting of such growth will lead to the formation of equilateral triangular pits, as observed.

It is of interest to note that this shallow pit is 400 A.U. deep. Pits ranging from some 60 to 600 A.U. deep (30–300 atom layers) have been measured on this particular surface.

Curvature of faces

Referring back to Figs. 46, 47, it is seen from the fringes that (1) growth sheets and their heights can be rapidly evaluated; (2) the majority of the growth sheets are curved. Again a controversial point is settled. Curvature of diamond faces is common and has been attributed variously to etch and to growth. The sheets evidently grow in stepped pyramid manner, as shown by a section illustrated in Fig. 48, and these when not resolved fully will lead to effective apparent curvature without postulating solution. There are irregular pits in Fig. 45 different from the rectilinear triangles. These irregular pits are most likely due to solution or etch. The curvature of the bases of growth triangles and of the tops of growth sheets, yet associated with rectilinear edges, affords proof that curvature and growth go together.

This particular diamond was removed and both the crystal surface and the flat resilvered at intervals. On setting up again the crystal would not be matched each time against exactly the same area on the flat. However, the interference pattern obtained remained unchanged, showing that the topography recorded is attributable to the diamond only.

It can be noticed that within some of the highly sensitive uniform tint regions there appear secondary fine-grain structure factors. These are of two kinds: (*a*) regular narrow fine-scale

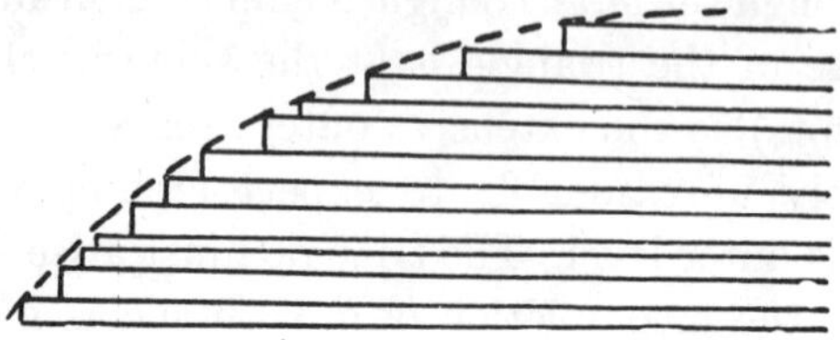

FIG. 48.

striations; (*b*) irregular 'honeycomb' patterns. The regular striae are undoubtedly crystallographic and characteristic of the diamond, for they appear within triangular areas and are crystallographically oriented, being parallel to a triangle edge. They therefore belong to the true topography.

The other 'honeycomb' details are more difficult to account for. They may be due to (1) the crystal, (2) the silver, (3) the flat. These honeycomb patterns look like diffraction effects and might be due to silver clumps acting as diffracting centres. With respect to this view it should be noted that such effects were not noticed on the fringes formed by doubly silvered mica. It has been shown now that the honeycomb patterns arise from fine scratch polish-marks on the optical flats used. This will be proved later in the sections dealing with white-light fringes. They do not therefore belong to any topographical features of the diamond surface.

One other irregularity may be mentioned since it is observed at times. A characteristic about some silver surfaces deposited upon crystals is that, unless the surface is very clean before deposition, the silver film begins to break up after some days and minute pin-holes appear in it. These become immediately noticeable in the photographs as strong black spots surrounded by diffraction haloes. They are quite characteristic and are a firm indication that resilvering is necessary.

A scrutiny of one of the crossed fringe interferograms shows in an interesting way that the uniform tint fringes are only sensitive when the separation is correctly chosen such that the light intensity is close to half the maximum. Why this should be so is illustrated in Fig. 49, in which, for clarity, an exaggerated fringe width has been drawn. A change in phase δ at A (due to a change

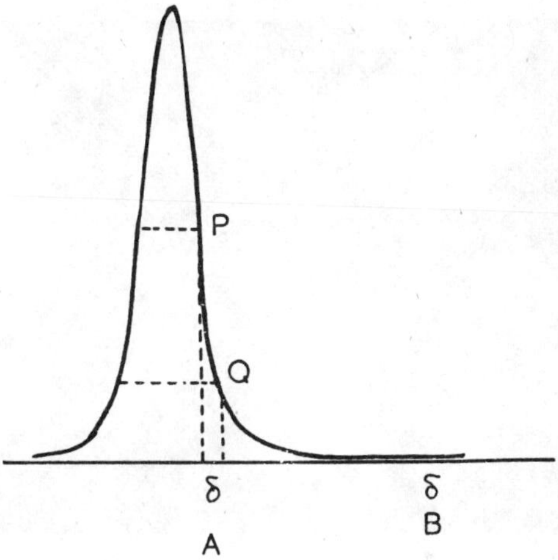

FIG. 49.

in thickness at the correct situation) leads to a fall of intensity represented by PQ. The same phase change at B has practically no effect on the very low intensity transmitted in this region.

Reference to the crossed fringe interferogram will reveal this in many regions. One can see, for example, in the light (correctly illuminated) regions in the section shown in Fig. 50, how, when a straight line fringe shows only a slight curvature, the whole field behind it may yet reveal a marked change in intensity. On the other hand, within the dark (insensitive) regions, the straight line fringes often display violent kinks and curvatures with no alteration in background illumination taking place, in accordance with Fig. 49.

It is for this reason that an unfiltered arc is used for the production of the high-dispersion fringes, for the different wavelengths help to fill in the empty regions and give sensitivity over a much bigger fraction of the field of view. Successful interferograms have been obtained using a special mercury arc containing added cadmium, thus increasing the number of lines in the field, yet not causing too much confusion.

So sensitive is the set-up when used for high dispersion, that photographic exposures must be kept to a minimum in order to reduce any possible minute disturbance due either to shock

FIG. 50.

or temperature change. The brilliance of the source can be heightened by raising the current, since the associated increase in line width is of little consequence.

A good achromatic (preferably apochromatic) objective is necessary if the completely unfiltered mercury arc is used, otherwise the blue and yellow patterns do not coincide and this causes confusion. For this reason a mercury-cadmium arc, with the blue filtered out, is likely to be superior in general to an unfiltered simple mercury source.

In making the crossed fringe exposure care must be taken not to over-expose (rather to under-expose) the high-dispersion pattern, for the light from this pattern photographically tends to broaden the sharp fringes later superposed on them. The procedure is simple when photographing with a microcamera which has an auxiliary eyepiece permitting the pattern to be viewed even whilst photographing. The silvered surfaces are set in the high-dispersion region and the camera shutter then opened. After completion of exposure, without touching the camera, the silvered surfaces are suitably tilted by a slight touch on an adjusting spring and the second exposure made, superposed on the first. The exposure times for the two systems are usually different and trial exposures require to be taken separately before the combination photograph is made.

FIG. 51.

Further studies on diamonds

It is clear from the crossed fringe interferograms already reproduced that each picture contains a mass of information. It is not the purpose of this discussion to consider individual crystals, but to illustrate principles and methods; hence the many interesting crystallographic features will not be reviewed here. Reproductions are shown in Figs. 51, 52 of crossed fringe

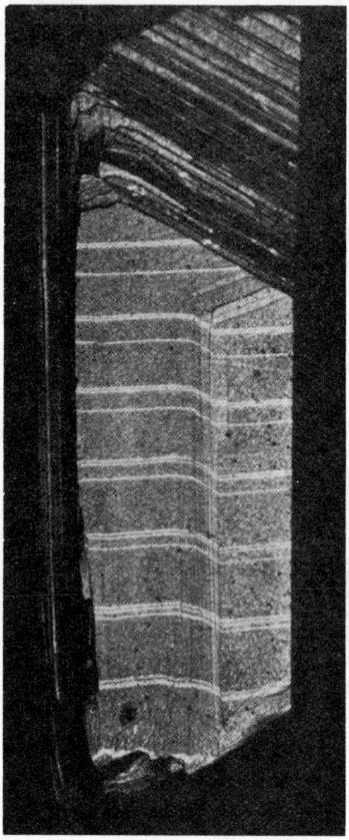

FIG. 52.

patterns given by two other diamonds (also portrait stones). The contrast in types is noticeable.

As illustrating the interesting new features which can be revealed, attention is drawn to the projections at the bases of the triangular growth markings shown in Fig. 51 the details of which are reviewed later. A considerable number of microphotographs of diamond surfaces have been variously published by many authors at different times. A scrutiny of a large number of these fails to reveal this peculiar feature on any of them. This does not imply that it does not exist; on the contrary it is an index of the sensitivity of the interference method for revealing sub-microscopic detail.

Comparison with the phase-contrast microscope

Within recent years there has been an important development in microscope technique, the invention of the Zernike phase-contrast method. Zernike has pointed out that microscopic fine detail can be regarded as being of the nature of either an amplitude-grating or a phase-grating. In an amplitude-grating the object consists of alternate transparent and opaque strips which are to be resolved, i.e. the *amplitude* of transmitted light varies as one progresses along the surface of the object. The phase-grating consists of a transparent body in which *phase* changes in the transmitted light are produced by different path lengths in the material under observation.

Zernike has shown that by changing the phase of the direct zero-order spectrum in the Abbe diffraction pattern, the contrast of detail in the image of such a transparent grating is much enhanced. As is now well known, he achieved this by introducing on his microscope objective a diaphragm of thin collodion, selected to produce the correct phase change. By this means detail-producing phase retardation is rendered visible.

However, it must be realized that this is in effect analogous to a two-beam interferometric procedure although the improvement is considerable.

The multiple-beam high-dispersion fringes can be considered as being equivalent to a super-phase difference microscope in that intense contrasts appear for minute phase changes. Very small raised features which are slightly visible in the Zernike microscope can and do exhibit most intense contrasts when viewed with high-dispersion multiple-beam interference fringes.

This explains how sub-microscopic detail can be revealed, and it is indeed possible that an extensive examination of other diamonds would reveal that features such as are recorded here are not exceptional and are possibly quite common.

Comparison with the electron microscope

It is of interest at this stage to contrast multiple-beam interference methods with the results given by the electron microscope. The first difference is, of course, that of cost and complexity. The complicated nature of the electron microscope is in striking contrast to the simplicity and inexpensiveness of the interferometric method. In a sense the two procedures are complementary. The optical method is probably the more sensitive. For the electron microscope essentially gives great magnification and high resolution over surface *area*, whilst the interference fringes give great magnification and resolution in *height and depth* only. The electron microscope can resolve, at considerable cost, some 50 A.U. The fringes, at very little cost, resolve 5 A.U. The electron microscope usually requires an elaborate technique for taking moulds of surfaces, whilst the optical method can often be directly applied. It is to be noted that by the adoption of an elaborate stereoscopic technique, usually produced by tilting the object, it has been claimed that, with the electron microscope, depth features of the order of 100 A.U. can be measured. Whether these claims are fully justified is not yet clear.

The actual linear magnifications produced by the high-dispersion white-light fringes to be described later (p. 96) exceed those usually reported for electron microscopes. Two fringes can be 10 cm. apart on a plate and this distance corresponds to $\lambda/2$, i.e. approximately 2,500 A.U. The linear magnification is then some $\times 400{,}000$. Far greater magnifications can be produced (hundreds of times) in special instances, as, for example, in the non-localized circular fringes (Chap. XV).

Polished diamond

The diamond interferograms discussed above were given by natural diamond surfaces which had only been subjected to cleaning processes by chemical reagents. The resistivity of diamond to chemical attack even after prolonged contact with acids and alkalis is well known.

It was of interest to examine a highly polished diamond face.

A diamond 'flat' was obtained and examined, the interferograms being as shown in Figs. 53, 54. The fringe structure is ragged, and in the high-dispersion region in Fig. 53 can be seen a regular succession of parallel streak markings. This plate shows fringes

FIG. 53.

with the green mercury line. These may be crystallographic or they may be simple polishing scratches. In any case their orientation and parallelism is such as to indicate that they are parallel to the 'softer' direction on the surface. Fig. 54 shows the yellow mercury fringes, the area being 2 sq. mm.

The surface is very different from that of a polished piece of glass, which gives evidence of local flow. A somewhat related effect has been observed in the comparison of the fringe quality given respectively by crystalline quartz and fused silica flats. The fused silica flats are locally smooth; the hard crystalline quartz shows a fine-grain structure.

Finally it may be noted that observations with diamonds are at times liable to suffer complication from ghost images. These

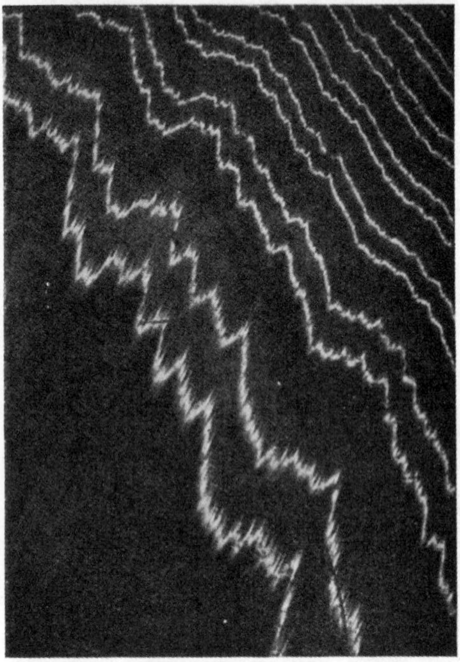

FIG. 54.

arise from the very high natural reflectivity of the unsilvered back face of the diamond. This has a reflectivity of 18 per cent., and from such a surface a ghost of appreciable intensity can be formed. If recognized the ghost introduces no particular complications. The ghost can be seen in Fig. 53. Polished diamonds are of much importance in instrument bearings, etc., and a more detailed study of such polished surfaces by interferometry should prove profitable.

CHAPTER VII

DOUBLY SILVERED FILMS: FIZEAU FRINGES

General properties

In the experiments previously described an optical flat was used as plane of reference. Clearly any method for avoiding this will prove of value, and in special instances this can be achieved. Thus it is easy to obtain thin slips of cleaved mica, and if both sides are silvered the phase condition will not be violated and sharp fringes will result, without the intermediary of a flat. Since the topographical features of any mica surface are always complex due to a combination of flexure and cleavage, it might have been anticipated that the Fizeau fringes formed with a doubly silvered slip would be still more complex because of superposition from both faces. This is, however, not the case in mica, for the reason given below.

Experiments can conveniently be made with thin slips varying from 1/50 to 1/150 mm. thick. Such sheets are flexible and sensitive to air shock, hence in most observations made the silvered pieces of mica are enclosed between glass plates held together with gummed paper strips. Such an arrangement leads to the formation of weak low-visibility secondary fringes formed in the air film between a glass surface and a silvered mica surface.

The doubly silvered mica is illuminated with parallel monochromatic light (green mercury) at normal incidence, using a small source at the principal focus of a good lens. To reduce the complexity arising from the birefringence of mica, plane polarized light can be used. The interference patterns which result are of interest. The visual appearance is still more striking if an unfiltered mercury arc is used as the source. Interference patterns (monochromatic light) for different mica samples are shown in Figs. 55, 56, each representing some 3 sq. cm. The sample shown in Fig. 55 exhibits moderately large areas of uniform tint separated by cleavage lines which radiate from the point of insertion of the needle used for separating the sheet. These cleavage lines can be seen on the mica surface with a low-power lens. The sample in Fig. 56 shows a mass of irregular torn cleavage strips of considerable complexity. Patterns between these two extremes have been observed in different samples and will be described later (p. 119).

FIG. 55.

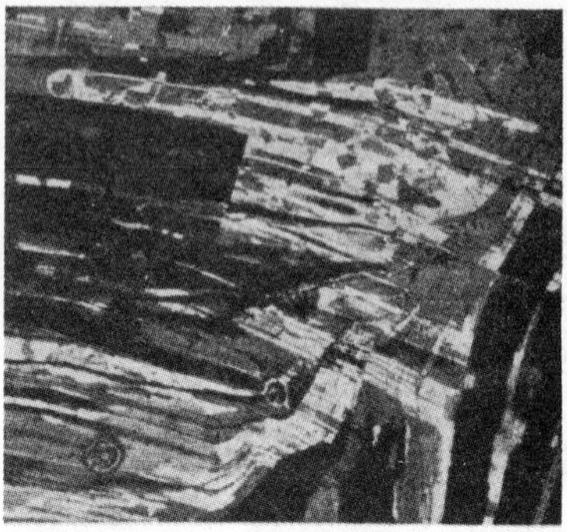

FIG. 56.

Within the uniform tint areas are three major characteristic features:

(*a*) Gradual changes in depth of tint.

(*b*) Striations and irregular discontinuous patches.

(*c*) Secondary low-visibility fringes.

The characters (*a*) and (*b*) are due to changes in refractive index arising from variations in chemical composition.† The low-visibility secondary fringes still appear when the mica rests upon an optical flat and thus represent the contour of the surface topography of that mica surface in contact with the glass.

In Fig. 57 the doubly silvered mica rests upon a *silvered* optical flat, and compound Fizeau fringes result. The mica acts as a filter leading to uniform tint areas bounded by cleavage lines. Within each area of transmission sharp multiple-beam fringes form between the silvered glass and mica surfaces. Uniform tint shows that both faces of the mica are parallel to within a small fraction of a wave, yet the secondary fringes show that the surface is highly contorted. It must be concluded that the distortions on the one face are faithfully followed on the other face

† See p. 93.

of the mica slip. Because of this the main features of the fringe pattern from a piece of doubly silvered mica are much less complex than those of either separate face when matched against a flat.

The 'parallel distortions' on the two faces do not prevent

FIG. 57.

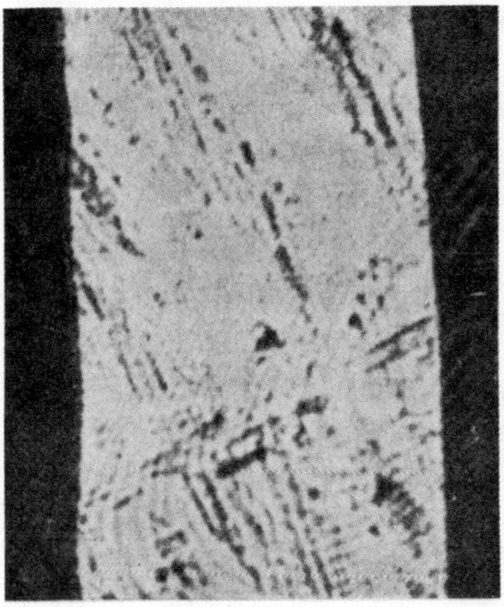

FIG. 58.

formation of uniform tint, which is determined by the condition $n\lambda = 2\mu t \cos\phi$. For t is postulated constant and the minute variations in ϕ due to the heights of the surface contours are negligible. Thus it is clear why good Fabry-Perot fringes (or the original Haidinger rings) can be formed by mica with very imperfect surfaces.

Fig. 58 is an enlarged section of fringes from a thin mica slip. Low-visibility secondaries can be seen, but in addition striated inclusions, clearly laid down in different levels, can be easily distinguished. A considerable amount of information about 'invisible' inclusions has been obtained by observations of this nature.

Owing to the local parallelism of opposing sides the Fizeau fringes fail to give information about the directions of steps and their values, merely indicating their existence, in a sensitive manner. It will be shown later, however, that more information can be obtained from doubly silvered films by employing other types of fringes in conjunction with the Fizeau fringes described here.

Wrinkling of the film

A matter of some significance is the effect of 'wrinkling' of the film. A local wrinkle leads to variation in ϕ in the expression $n\lambda = 2\mu t \cos\phi$. As in the case of the collimation error discussed in Chapter II, a change from normal incidence to an incidence ϕ produces an order change dn, where

$$dn = \frac{4t}{\lambda}\sin^2\frac{\phi}{2} = 2n\sin^2\frac{\phi}{2}.$$

From this a table can be constructed giving the change dn produced by a range of ϕ values, for various t values, as below. For approximate calculation λ is taken to be 5×10^{-5} cm. and only approximate dn values are given.

TABLE IV

t mm.	*1*	*0·1*	*0·01*	*0·001*
$\phi°$	dn	dn	dn	dn
0·2°	0·024	0·002	0·0002	0·00002
0·5°	0·15	0·015	0·0015	0·00015
1·0°	0·6	0·06	0·006	0·0006
2·0°	2·4	0·24	0·024	0·0024
5·0°	15	1·5	0·15	0·015

Table IV shows that the value of t determines the angle of wrinkling that can be tolerated. Taking $t = 0{\cdot}01$ mm., which is the upper limit determined by the phase condition, angles up to $\frac{1}{2}°$ can be tolerated since they produce a displacement of no more than 1/600th of an order. As the square of the angle is involved the effect increases rapidly with ϕ and for 2° *dn* is 1/40th of an order, a quantity no longer negligible, whilst for 5° *dn* is as much as 1/6th of an order.

This effect must be carefully guarded against in special cases. For example, it is possible to silver very thin films of plastics and to examine them by the multiple-beam technique. If the film wrinkles severely, then this factor must be taken into account in the interpretation of the observations. The wrinkle angles that can be tolerated without noticeable disturbance are of the same order as the defect angles permitted in collimation.

It will be noted that a *severely* wrinkled film, even of uniform thickness, leads to a fringe pattern which appears to give a spurious structure if the wrinkling be not recognized.

In general the detailed interpretation of doubly silvered films can only be made with the aid of the white-light fringes described in the next chapter, yet general information about structural defects, etc., of thin films can be obtained in a qualitative manner from the Fizeau fringes alone, as indicated in this chapter.

CHAPTER VIII

FRINGES OF EQUAL CHROMATIC ORDER

Origin of fringes

IN the preceding chapters the discussion has been restricted to fringes produced by one or at the most a few separate monochromatic radiations. It will now be shown that considerable advantages accrue in particular investigations by the use of a different type of interference method employing white light. These white-light fringes will be called *fringes of equal chromatic order* for reasons given on pages 97–9. The precision and variety in range of application of these fringes probably surpasses that shown by other types of low-order interference.

A notable feature is the optical simplicity of the set-up and the fact of attainment of high precision without a monochromatic source. Despite the use of white light, the fringes obtainable are sharper than those given by monochromatic Fizeau fringe methods, sharper in the sense that the fringe width occupies a smaller fraction of the distance between orders.

As before, the interference arises initially within a thin film bounded by silvered surfaces. Since very bright white-light sources are readily available (e.g. high-current carbon arcs) higher reflecting coefficients can be adopted than can be used with monochromatic methods, and also higher magnifications can be employed. It is possible to use $R = 0{\cdot}94$ with $\times 400$ magnification and yet obtain strong exposures in a few minutes.

The expression controlling the fringe shape is the Airy formula

$$I = \frac{I_{\text{max}}}{1+F\sin^2\pi\{(2\mu t/\lambda)\cos\phi\}}.$$

If μ is taken to be unity the variables in this are t, ϕ, λ. This leads to four types of fringe as stated in Table V. Provided the conditions permit a close approximation to Airy summation then, whether the $\sin^2$ term variation is produced either by t, ϕ, or λ, or combinations of these, the fringe-intensity distribution will be the same.

TABLE V

Fringe types

Nature of light	*Constant quantity*	*Fringe type*	*Name*	*Filter action*
Monochromatic (λ constant)	t	Equal inclination	Fabry-Perot	Angular
	ϕ	Equal thickness	Fizeau	Linear
White (λ variable)	ϕ	Equal t/λ	Equal chromatic order	Wave-length
	t	Equal $(t\cos\phi)/\lambda$	White-light Fabry-Perot	

With monochromatic light (λ constant) the fringes with t constant are called 'fringes of equal inclination', the Fabry-Perot fringes being typical. With t variable and ϕ constant, 'fringes of equal thickness' arise, wedge Fizeau fringes being typical. A Fabry-Perot interferometer is thus in effect an *angular filter*, only passing light incident at those angles ϕ for which n is integral in the expression $n\lambda = 2t\cos\phi$. A multiple-beam Fizeau wedge set-up behaves as a *linear filter*, light being passed at points along the wedge surface for those values of t satisfying the above expression (generally making $\cos\phi = 1$ also).

With λ variable, if ϕ is constant and t varies, fringes are produced in which t/λ is constant for each fringe, since this represents the order of interference. Clearly any particular value of t passes a number of discrete wave-lengths for which n is integral. Thus if t is constant (plane parallel film) the system can be regarded as a *wave-length filter*, passing a group of discrete wave-lengths. If t varies from point to point, then each point passes its own wave-length group. If then an image of the film is projected on to the slit of a spectrograph, each point on the slit passes a group of wave-lengths corresponding to the local value of t. The integration along the slit leads to fringes in which t/λ is constant (order of interference), and as these fringes cover a wave-length range (often wide) they are chromatic, hence the name *fringes of equal chromatic order* is proposed for them [26].

Finally with λ variable, t constant, and ϕ variable the fringes formed are 'white-light Fabry-Perot fringes' [27], but since these have little bearing on the problems under review, they will not be discussed here.

Optical arrangement

The restrictions concerning both collimation and phase condition are the same as for the case of multiple-beam Fizeau fringes. The fringe dispersion in Fizeau fringes depends upon

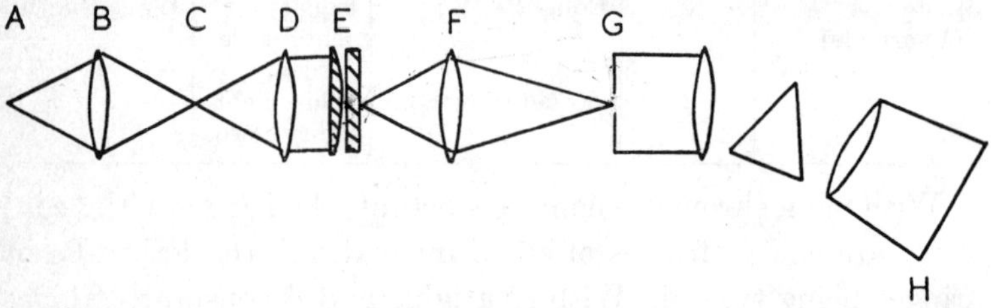

FIG. 59.

the wedge-angle θ, and if a high magnification is to be used, then θ must be made sufficiently large to retain a reasonable number of fringes in the field of view. This increase in θ tends to make the phase-retardation effect more serious and ultimately fixes a limit to the Fizeau fringe sharpness. Indeed, very high magnification (say $\times 500$ or more) is likely to be difficult on these grounds. It is shown on page 100 that in fringes of equal chromatic order the *value of θ has no influence on the dispersion.* The latter is determined only by t and the spectrograph used. It is thus possible to reduce θ until the surfaces are as near to parallelism as is possible. Thus the optimum condition for reducing the phase lag exists. For this reason, and because higher reflecting coefficients can be used, fringes of equal chromatic order can be produced which are inherently sharper than the narrowest Fizeau fringes yet obtained. Furthermore, high magnifications are employable with no reduction in sharpness, which is *not* the case with Fizeau fringes.

To consider the optical arrangement, the fringes formed by using the silver-modified Newton's ring lens-plate combination as the interference film will first be described. The arrangement used is shown in Fig. 59.

A is a white-light source, an image of which is projected by a lens B on to a circular aperture C (some 2 mm. in diameter, the value not being critical). C is at the focus of an achromat lens D, which projects a parallel beam of white light on to the interference system E, the incidence being normal (if non-normal incidence is used the phenomena are complicated by the differential polarization phase-change doubling).

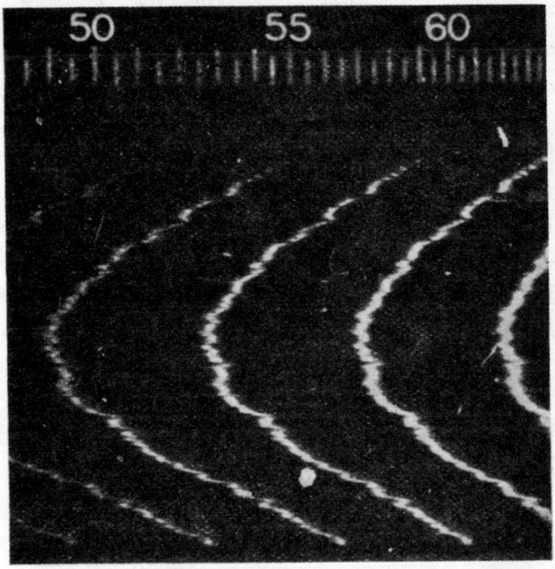

FIG. 60.

With the good achromat lens F or a microscope objective an image of the film E is projected on to G, the slit of a prism spectrograph. In the experiments described below this was a medium quartz (Hilger) spectrograph. The fringes appear at H, where they can be observed with an eyepiece or photographed. A reproduction of the fringes is shown in Fig. 60, which covers part of the visible region from the red to the green. A wavelength scale is superposed. These fringes correspond to the narrow section of the interference film E, which has been selected by the slit G, since an image of E is projected on to G.

Attention may be drawn to the following special features:

(1) The separation between fringes, in Ångström units, is considerable.

(2) The fringes are sharp and narrow, occupying but a small fraction of the wave-length interval separating a fringe pair.
(3) Each fringe exhibits a fine structure.
(4) Each fringe is similar to its neighbours.
(5) Each fringe is multicoloured and can extend over some thousands of Ångströms, according to the magnification of the lens F.

Before discussing the particular fringe-shape shown in Fig. 60 (which is determined by the geometry of the particular interference film) some general properties applicable to any type of thin film will first be considered.

Each of the curved fringes belongs to the third system given in Table V since for any fringe t/λ is constant. Each point on a given fringe corresponds to a different path length within the film, yet containing the *same number of waves*, for the wave-length varies regularly from point to point to compensate for the change in path. The expression $n\lambda = 2\mu t \cos\phi$ is simplified here by writing $\mu = 1$ and $\cos\phi = 1$, for the incidence is restricted to normal. It can be rewritten

$$n = 2\nu t,$$

where ν is the wave-number.

Differentiating gives $dn = 2d\nu . t$.

For successive orders $dn = 1$ and writing the wave-number separation between orders as $\Delta\nu$ gives $t = 1/2\Delta\nu$. The fringes are thus equally spaced for any fixed value of t. A useful extension is to write $t = m/2\Delta\nu_m$, where $\Delta\nu_m$ is the wave-number interval between $(m+1)$ fringes.

The shape of the fringes formed with a lens-plate combination

In the lens-plate one can imagine white-light Newton's rings being formed in the film, then being projected on to the slit. The spectroscope sorts out the mixed fringes, which are generally not visible as such on the slit, except possibly very close to the centre. Thus $\rho^2 = 2R(t-\epsilon)$, in which R is the radius of curvature of the lens face, ρ the radius of the Newton's ring formed at a film thickness t for a wave-length λ, ϵ being the *optical* separation at the 'point of contact' of the lens and plate. The quantity ρ must be corrected for any enlargement introduced by the projecting lens.

Let λ_0 be the wave-length on a given fringe at a point on the slit on which falls the image of a point on the film at which the thickness is t_0. Let λ_a be the wave-length position of the same fringe at a further position on the slit image away from the centre, corresponding to a film thickness t_a.

Then:
$$n_0\lambda_0 = 2t_0,$$
$$n_a\lambda_a = 2t_a,$$
$$n_a\lambda_a - n_0\lambda_0 = 2(t_a - t_0) = (\rho_a^2 - \rho_0^2)/R,$$
where ρ_a, ρ_0 are the Newton's ring radii at the film thicknesses t_a and t_0.

At the centre of the fringe system the film thickness is ϵ, and if t_0 in the above refers to the centre, then $n_0\lambda_0 = 2\epsilon$ giving
$$n_a\lambda_a = (\rho_a^2/R) + 2\epsilon.$$

Since a fringe is the locus of points having the same value of n, then n_a is constant within a fringe. As ρ_a is the distance from the fringe centre, it follows that each fringe has a parabolic shape, provided the dispersion of the spectrograph is normal. With a prism instrument the parabola is distorted because of variable dispersion and there is a gradual change in shape on moving from the red to the blue.

It is clear that the shape is independent of the value of n (variable dispersion being disregarded), hence each fringe resembles its neighbours, even as to fine detail.

Some of the applications of the approximately parabolic fringes will now be reviewed. The following features can all be calculated from a single photograph.

(1) *Optical separation at the 'point of contact'*

The optical separation of the two silvered components at the 'point of contact' is obtained with high precision. It is $1/2\Delta\nu$, where $\Delta\nu$ is the wave-number separation between a pair of fringes at the centre of the pattern. This quantity can be evaluated with certainty to within 20 A.U., i.e. 1/250th of a light-wave (λ 5,000 A.). In Fig. 60 this separation is 37,040 A.U., i.e. some 6·78 light-waves (green mercury). This optical value includes phase change at reflection. The precision in evaluation exceeds that attainable by other methods. The extrapolation procedure usually advocated in text-books is not valid to this degree of precision owing to the quite unreliable assumption of true sphericity of the lens surface.

(2) *Variation of phase change with wave-length on reflection*

According to the simple theory given above, the value of $\Delta\nu$ should be constant for any pair of fringes. Measurement shows the existence of a quite small but regular increase in $\Delta\nu$ with diminishing wave-length. This can be attributed to the known small change in phase at reflection at the metal surface when the wave-length is altered. A method is thus available for the determination of this variation of phase change with wave-length.

(3) *The radius of curvature of the lens*

The curvature of the lens face can be obtained from the fringes by at least four independent methods of measurement:

(*a*) From the horizontal and vertical coordinates of any two points on a single fringe, but this requires also a knowledge of the dispersion of the spectrograph.

(*b*) By measuring at two horizontal levels, a known distance apart, the number of fringes (including fractions) between two specific wave-lengths.

(*c*) By evaluating $\Delta\nu$ at different horizontal levels (two suffice) from one pair of fringes. Both this method and the preceding give the thickness of the air gaps at two points, the effect of phase change at reflection being eliminated.

(*d*) A more general procedure, which further reveals any departures from true sphericity, is to measure the positions of the intersections of the fringes by any vertical line. This gives the effective Newton's ring diameters for the particular wave-length selected by the vertical section.

The irregularities in the fringes

A feature of considerable value is revealed by the irregular fine-detail structure of the fringes. These details have their origin in the small-scale irregularities on the glass surfaces of the lens and plate. The plate employed was not an optical flat, but the irregularities do not entirely vanish if a high-grade optical flat is used. The residual features are then the small-scale surface errors on the lens face. These errors are shown variously either by the wrinkles in the circular monochromatic fringes of Fig. 18 or by the short zigzag irregularities in the fringes of equal chromatic order in Fig. 60.

The dimensions of these small-scale features can be calculated in a simple manner. Since $n\lambda = 2t$, then $dt = dn\,.\,\frac{1}{2}\lambda$. It is therefore necessary only to determine, at any selected wave-length, the fraction of an order by which the small feature is displaced from neighbouring positions in order to obtain the change in thickness dt responsible for the production of the irregularity. Since the order separation is linear, in terms of wave-numbers, and practically linear in terms of wave-lengths if $\Delta\nu$ is not large, the fraction is obtained readily.

Fringes can be produced which are so sharp that a $dn = 1/250$ can be measured, which corresponds to only 10 A.U. This quantity is even smaller than the crystal lattice spacing of several well-known crystals. It is this striking optical achievement that makes the fringes so useful. It will be shown later that the magnification produced by the projecting lens F has an important effect upon the fringe definition.

CHAPTER IX

APPLICATIONS OF FRINGES OF EQUAL CHROMATIC ORDER

I. MICA SURFACES

Mica topography

THE fringes of equal chromatic order have numerous possible applications, and the examination of the surface topographical features of a mica sheet will indicate some of their properties. The silvered mica is pressed against a silvered flat, as before, and replaces the Newton's ring lens-plate combination. In the Fizeau method described on pages 54–64 it was not easy to distinguish hill from valley, nor could the true direction of the step pattern be decided.

Such ambiguities are resolved by employment of the fringes of equal chromatic order. Furthermore, the precision of measurement in the latter is not affected by local variations in dispersion which is so marked a restricting feature of Fizeau fringes. It is clear that the spectroscope slit selects only a narrow section of the mica surface and reveals its topographical features. To cover any appreciable area of the mica it is necessary to traverse the mica image across the slit by moving the mica horizontally across the field. Whilst this is of necessity a laborious procedure photographically, it is a rapid procedure visually, and a survey of considerable mica areas can quickly be made.

A typical fringe pattern is that shown in Fig. 61, the lower picture being an enlarged section of the upper. Reference back to the fringes given by the plate-lens show which regions of the mica represent hills and which valleys. Fringes convex to the violet represent hills, and those convex to the red valleys. Further, when the number of fringes in the field is small, this represents a high spot, and so the contour can readily be visualized. Not only is the direction of incline shown, but by measuring the order values at the peaks, planes, and valleys, precision values of the heights and depths are obtained.

From the consideration of any vertical section the rate of change of height, i.e. the steepness of the incline, is obvious.

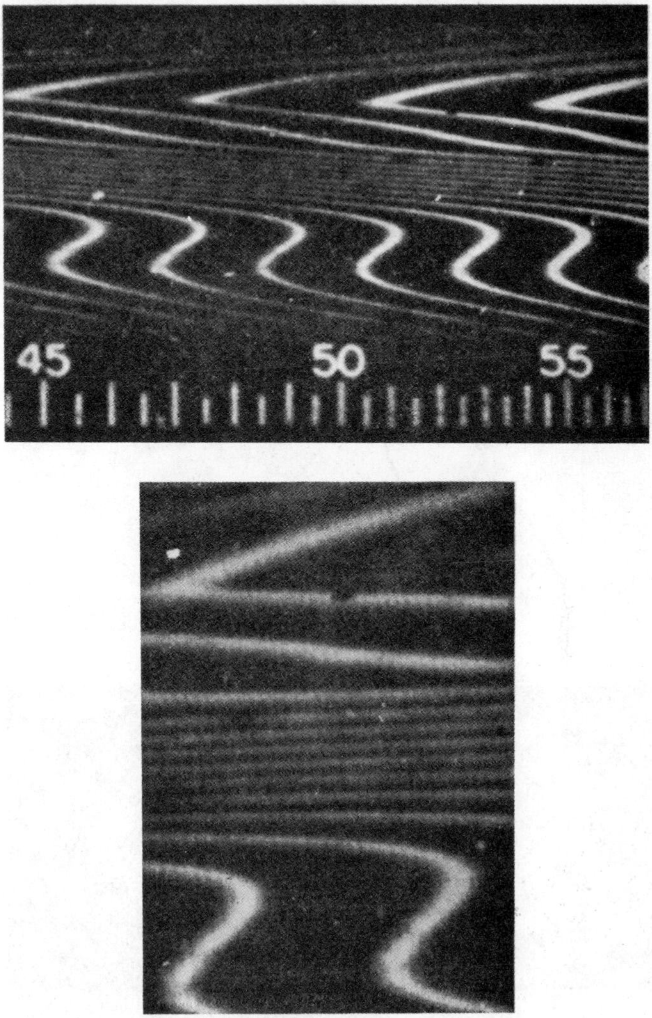

FIG. 61.

The number of fringes intersected per unit of length of the vertical section is a measure of the surface slope in terms of half-waves. The more close-packed the fringes the steeper the slope.

The detailed analysis of this pattern is shown graphically in Fig. 62.

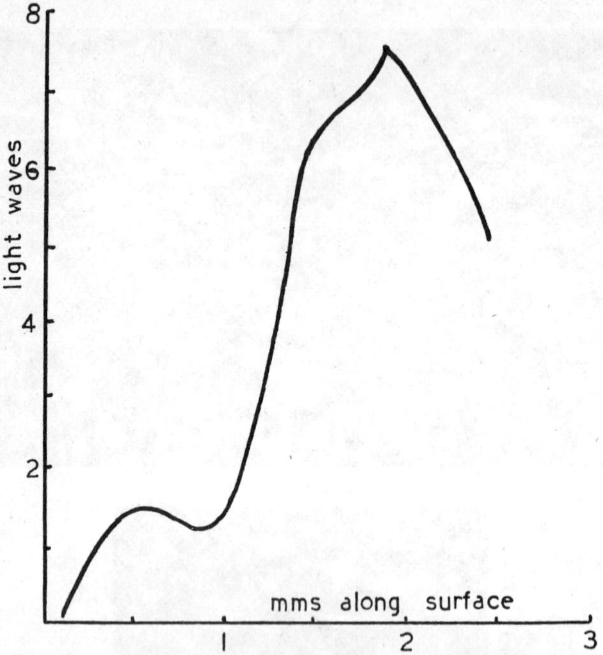

FIG. 62.

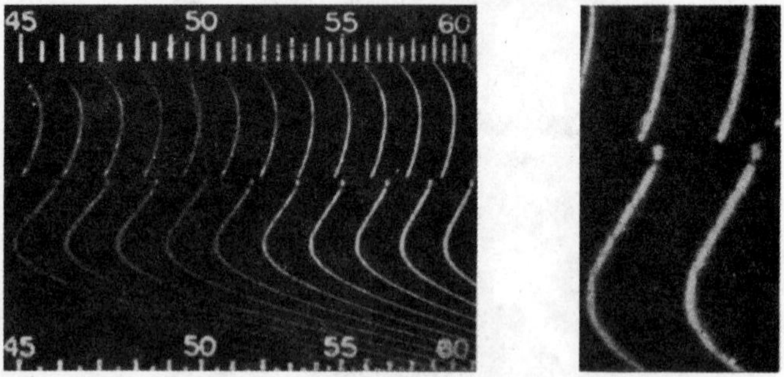

FIG. 63.

An interesting effect is shown in Fig. 63. In this a mica specimen has been placed such that a surface cleavage step crosses the slit image. A discontinuity thus appears. Two separate conclusions can be drawn from this record. From the difference in the number of fringes, in a given wave-length range, on either side of the discontinuity, the height of the step can be calculated. Further, the side containing the larger number of fringes is farthest from the optical flat, hence the direction

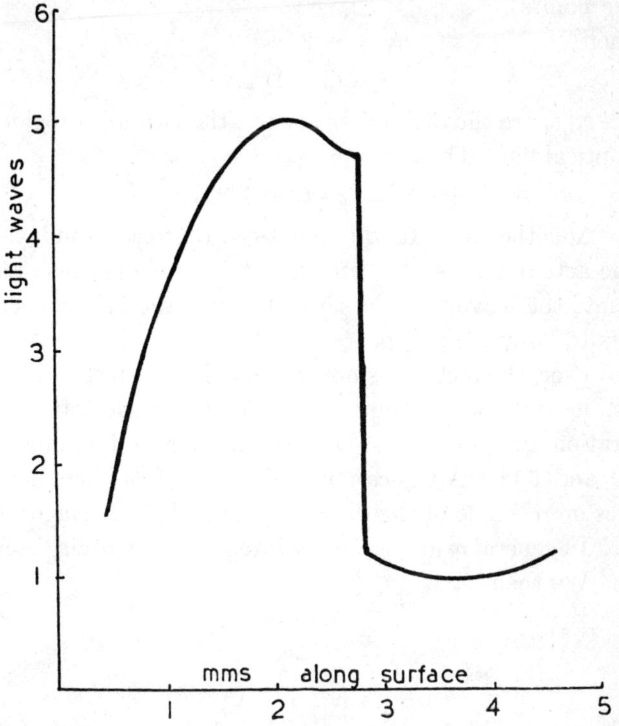

FIG. 64.

of the step is clear by inspection alone. The result of detailed measurement is given in Fig. 64.

The following treatment can be employed for the evaluation of a step. The procedure will be called the method of coincidences.

Coincidence method for evaluation of cleavage steps

It will be seen that the fringes in Fig. 63 exhibit either perfect or imperfect coincidences as in the case of a typical vernier scale. The step can be calculated either from perfect or imperfect coincidences in a simple way.

Consider first a perfect coincidence.

Let there be $m+1$ fringes in one part of the field and suppose these coincide with $m+2$ fringes in the other part. Let $\Delta\nu$ be the frequency wave-number difference between the two coincidence points.

Then
$$t_1 = m/2\Delta\nu,$$
$$t_2 = (m+1)/2\Delta\nu,$$
where t_1, t_2 are the distances between the two mica regions and the optical flat. Thus
$$\Delta t = t_2 - t_1 = 1/2\Delta\nu,$$
where Δt is the step. It will be noticed that this is independent of the actual values of t_1 and t_2. It is thus only necessary to evaluate the wave-number separation between the coincidence points to obtain the step.

However, the method is not restricted to perfect coincidence. Thus, let two wave-lengths be selected, with wave-number separation $\Delta\nu$ (for ease in computation round numbers, say 4,000 and 5,000 A.U., can be selected). Let there be $m+1$ fringes on one side of the dividing line and $n+1$ fringes on the other. In general m and n are not integers but contain fractional parts. We then have
$$t_1 = m/2\Delta\nu,$$
$$t_2 = n/2\Delta\nu,$$
$$\Delta t = t_2 - t_1 = (n-m)/2\Delta\nu.$$
If it happens that Δt is a fraction of a half-wave, then for convenience $\Delta\nu$ can be chosen so that n and m both contain the same integer and can be written $n+x$, $n+y$, respectively, where x and y are fractions of an order. From this $\Delta t = (x-y)/2\Delta\nu$. It is thus only necessary to measure the fractional parts x, y at the ends of the selected wave-length range.

There is a still simpler procedure than the use of coincidences if Δt is a *small* fraction of a half-wave. For then the relation $\Delta t = dn . \frac{1}{2}\lambda$ can be used. It is only necessary in this case to measure the fraction of an order displacement dn at *one* wave-length. A displacement to the violet corresponds to a step up towards the optical flat.

The measurement of small angles

The fringes of equal chromatic order can be used as a particularly sensitive device for the measurement of small surface angles. For example, ridges on surfaces which depart only slightly from a true plane can be studied, a characteristic example being shown in Figs. 61, 62 which refer to mica. In particular, slightly inclined vicinal faces on crystals (i.e. faces with large Millerian indices) can be detected. The sharp, acute angle between the almost straight fringes in Fig. 61 represents a small surface ridge.

A wedge between planes, placed with the edge perpendicular to the spectroscope slit, leads to the formation of straight-line fringes sloping across the field of view. From the measured fringe slope the wedge angle can be calculated. Let the horizontal distance between two orders be x cm. on the photographic plate and let y be the vertical distance between two fringes (the fringes are parallel). The measured fringe slope is x/y. But the true wedge angle is $\lambda/2y$. The magnification is then $2x/\lambda$. With the spectrograph used the dispersion at 5,000 A. is some 200 A./cm. Taking $x = 5$ mm. between orders leads to a value of 20,000 for the magnification of the angle. This argument implies unit magnification by the lens.

The angles which are thus measurable are much smaller than the Rayleigh limit. According to the accepted Rayleigh procedure the smallest angle measurable by a goniometer is λ/y, where y is the length of the wedge. But with these highly sharpened fringes a displacement of $\lambda/200$ is measurable at a discontinuity, and in the case of a continuous sloping line $\lambda/50y$ can be evaluated, and under suitable conditions $\lambda/100y$. Angles 100 times smaller than the Rayleigh limit can therefore be measured with certainty.

The magnification of the projecting lens

The magnification produced by the projecting lens has an important bearing on the experimental attainment of the maximum resolution, in accordance with the following considerations. In the case of the lens-plate combination (Newton's rings) the projection of an enlarged image leads to appreciable sharpening in the fringe definition. For, by enlarging, the spectroscope slit is in effect selecting a portion of the interference film, narrower than itself, the region under observation being reduced in width by the magnification ratio. So sensitive is the experimental procedure to small changes in thickness that the minute variations introduced across even a narrow slit width can at times lead to a considerable reduction in ultimate sharpness, unless magnification is employed to counteract this.

We have here the paradoxical optical situation in which the enlargement of a projected image leads to an ultimate improvement in definition. More strictly, of course, the result of enlarging is geometrically equivalent to the reduction of a stop before the surface.

When the fringes are formed by a parallel-sided film, then magnification has no appreciable effect. A step, as in Fig. 63, is quite independent of this and fringe definition is also not effectively improved. If, however, sloping fringes are seen due to a wedge film, as, for example, in Fig. 61, then a diminution in the size of the projected image leads to an increase in the slope of the recorded fringes, and vice versa if an enlarged image is used.

The effects which occur are summarized by considering a wedge between two planes as the interference film. When the wedge is set with its edge parallel to the slit, magnification improves definition of the vertical fringes which form. When the wedge is rotated through 90°, being set with its edge perpendicular to the slit, magnification has no influence on the definition but reduces the effective slope of the fringes.

If a slope is suspected, the true wedge value is only obtained if the film is rotated to give the maximum fringe slope in the field of view, for a wedge set at an angle other than with its edge perpendicular to the slit gives an apparent wedge which is only a resolved component of the true wedge value. It is clear that a judicious employment of magnification must be adopted according to the problem in hand. The guiding criterion is that only the surface dimensions and not the film thickness are affected by the magnifying lens.

Alternative method of calculation

The simple relation $dt = dn \, . \tfrac{1}{2}\lambda$ is applicable only if dn is a small fraction of an order, or alternatively, if t is such that there are many fringes in the field of view. It is frequently desirable to reduce t to give high dispersion such that possibly no more than three or four fringes appear in the visible region. In this case it is necessary to determine the order of interference n, as follows. The order is clearly higher for the blue end of the spectrum. Suppose fringes are displaced, the displacement being less than one order; since $n\lambda = 2t$,

$$dt = \frac{n}{2} d\lambda$$

where $d\lambda$ is the wave-length change produced by dt, *which need not necessarily be small.* The quantity n is obtainable since

$$n\lambda = (n+1)\lambda_1$$

for two adjacent orders, giving

$$n = \frac{\lambda_1}{\lambda - \lambda_1},$$

with similar expressions for any other pairs of orders. Since n is a small number (in some observations $n = 6$ has been used) it can always be determined without ambiguity when high dispersion is used. Since both n and $d\lambda$ are now determined these give dt.

II. OBSERVATIONS ON THIN MICA SHEETS

Correlation of fringes

It has previously been pointed out that the uniform-tint Fizeau fringes given by thin mica slips, whilst revealing a good deal of *qualitative* detail, do not give much numerical information. The combination of Fizeau fringes and fringes of equal chromatic order is, on the other hand, a powerful arrangement. Fig. 65 shows a Fizeau fringe picture, and Fig. 66 the fringes of equal chromatic order corresponding to a line section 1·5 mm. in length crossing one of the cleavage discontinuities. The straight fringes, parallel to the slit, indicate that the mica faces are parallel. The fringes are double because mica is birefringent, and a polaroid demonstrates that the two components of each pair are plane polarized, mutually perpendicularly. They differ slightly in intensity because of differential absorption in the mica.

FIG. 65.

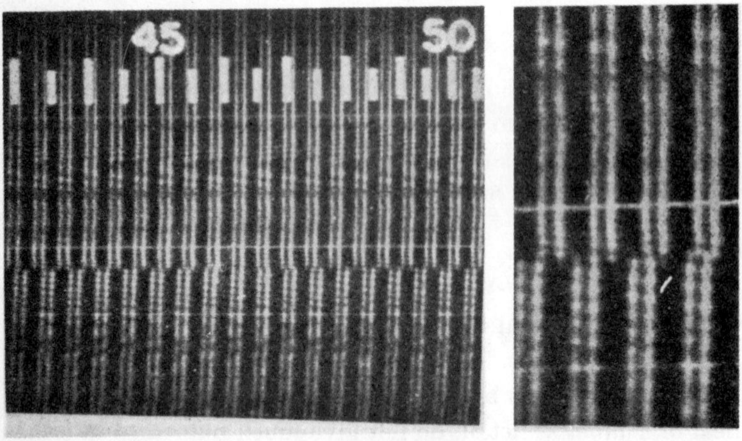

FIG. 66.

These fringes reveal many properties, e.g.

(*a*) The direction of the cleavage step.

(*b*) The height of the step. (It is 120 A.U., i.e. exactly 6 'molecules'.)

(*c*) The thickness of the mica. (It is 1/59 mm.)

(*d*) Secondary intensity variations superposed on the whole pattern.

(*e*) The birefringence over an extended wave-length range.

The secondary fringes reveal the surface topography, since they are the low-visibility fringes of equal chromatic order formed between the mica surface and the *unsilvered* flat on

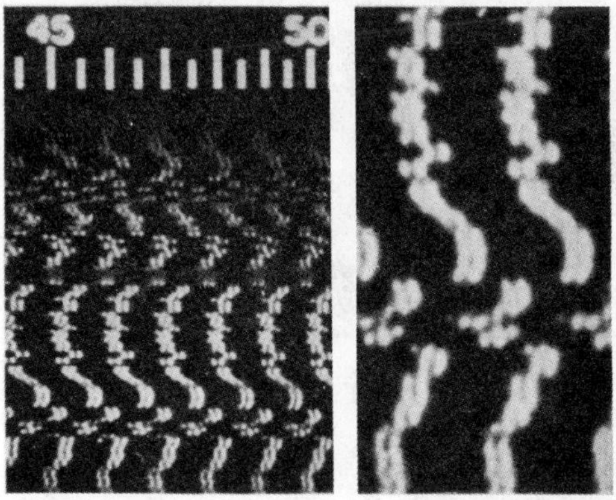

FIG. 67.

which this rests. This again proves that although the surface is contorted, yet the parallelism is retained. There exists in fact point-to-point correspondence between the Fizeau fringes and the fringes of equal chromatic order.

The fringes of equal chromatic order shown in Fig. 67 (enlarged section at right) refer to a 1 cm. length of part of the mica sample shown in Fig. 56. Again, correspondence can be traced, but it is clear that the fringes of equal chromatic order reveal much additional detail about the cleavage steps. One such conclusion, for example, is that changes in thickness are generally discontinuous (excluding some special local features associated with inclusions and growth nuclei).

A special characteristic of doubly silvered mica fringes is that, the 'optical' thickness being involved, displacements are magnified by the factor μ. The sensitivity is thus some 60 per cent. better than when using an air film as in the topographical studies on single surfaces matched against a flat. Fig. 68 exhibits a specimen with complexity intermediate between those of Figs. 66, 67.

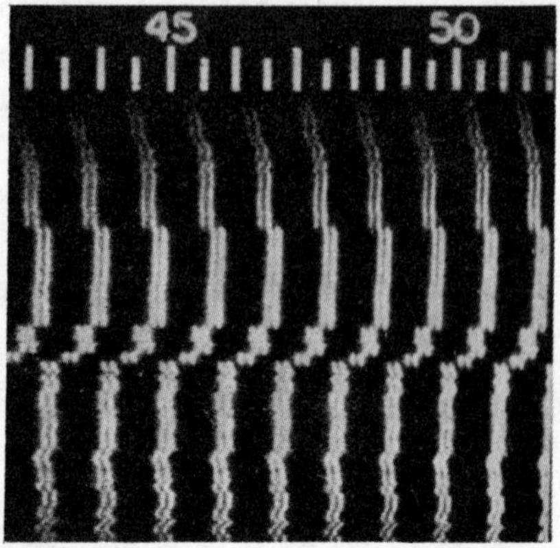

FIG. 68.

Birefringence of the mica

The fringes of equal chromatic order offer a new method for the measurement of birefringence and have the added advantage of giving the birefringence over the whole visible spectrum with a single photograph, provided the refractive index is known for one wave-length, which is often the case.

Strictly speaking the formula $n\lambda = 2\mu t \cos\phi$ when applied to a dispersive medium (i.e. $\mu > 1$) leads to somewhat more complex expressions for the separation between orders than is the case when $\mu = 1$. The resulting formula is *not* $\mu t = 1/2\Delta\nu$ but can be shown to be

$$\mu t = \frac{1}{2\Delta\nu\{1-(\lambda/\mu)(d\mu/d\lambda)\}}.$$

For a non-dispersive medium ($\mu = 1$) this reduces to the previous simple case of $t = 1/2\Delta\nu$. The dispersing term $(\lambda/\mu)(d\mu/d\lambda)$ leads to a correction of the order of 2 per cent. in a typical dispersive medium and for many measurements can be disregarded, permitting the simpler, approximate expression to be used.

The refractive indices of micas differ, the *averages* for the D line given in the International Critical Tables being 1·561, 1·590, 1·594. The expected birefringence is thus 0·004 or thereabouts.

Consider the wave-length change $d\lambda$ for a fringe of given order due to change in refractive index $d\mu$, we have:

$$n\lambda = 2\mu t,$$

$$n\,d\lambda = 2\,d\mu\,t,$$

i.e.

$$d\lambda = d\mu \,.\, \frac{\lambda}{\mu}.$$

By measuring $d\lambda$ the birefringence $d\mu$ is obtained if μ is known, say from tables. The value of μ can also be obtained from the fringes in the region of λ, thus:

Since

$$n = 2\nu\mu t,$$

$$dn = 2\mu t\,d\nu + 2\nu t\,d\mu,$$

and writing $dn = 1$, for the separation $\Delta\nu$ between orders, leads to

$$\frac{1}{2\mu t} = \Delta\nu + \frac{\nu}{\mu}\Delta\mu,$$

in which $\Delta\mu$ is the change of μ over the range $\Delta\nu$. Since in any case the birefringence is small, it suffices to use the approximate value of μ given by neglecting the term $(\nu/\mu)\Delta\mu$.

This approximate value of μ can then be used in the above expression $d\lambda = d\mu(\lambda/\mu)$ to give the birefringence $d\mu$. Alternatively, of course, the approximate value of μ can frequently be obtained from tables.

With the particular example illustrated in Fig. 66 the birefringence, $d\mu$, is sensibly constant over the visible region, the value at 5,000 A.U. being 0·0046±0·0001. It is to be noted that a birefringence of only 0·0001 could be measured. This would not lead to a resolvable doublet, but if the incident light were first polarized and then an independent wave-length determination made of the two separate components produced by rotating the polarizer, the wave-length shift for a birefringence of 0·0001 could readily be detected.

It is to be particularly noted that the birefringent wave-length doubling at any fixed wave-length,

$$d\lambda = d\mu \, . \frac{\lambda}{\mu} .$$

is independent of the thickness of the mica. This has been experimentally confirmed. We have here the unusual feature of an interference phenomenon which is *not dependent on the thickness of the interference film.*

Further applications

Although only the birefringence of mica has been discussed, it is quite clear that the method described has potentialities. An important possible experiment is as follows. A glass surface can be silvered and on this monomolecular layers of organic compounds including, for example, protein layers can be deposited. Silver can be evaporated on to the layer and the thin film will then reveal the existence of any molecular birefringence. Another possibility is the examination of the occurrence of any birefringence in lubricating oils by enclosing a thin oil film between silvered surfaces and applying forces.

The study of inclusions

A further example of the advantages gained from combining the Fizeau fringes with fringes of equal chromatic order is illustrated by considering their applications to the examination of the small inclusions frequently found in mica. Fig. 69 represents Fizeau fringes ($\times 80$) for a small local feature. The fringes are double, owing to birefringence. It is not possible from this photograph to decide which are the hill and which the valley features, nor is much detail observable. Fig. 70 shows the fringes of equal chromatic order taken over an approximately central section. The details of structure are immediately obvious. The central feature is a hill flanked by valleys. There is reason to believe that this particular pattern arises from the inclusion of a small transparent solid producing the hill. On either side, owing to incompletion of the mica growth layers are probably vacuoles containing either gas or liquid and these optically produce the 'valleys'.

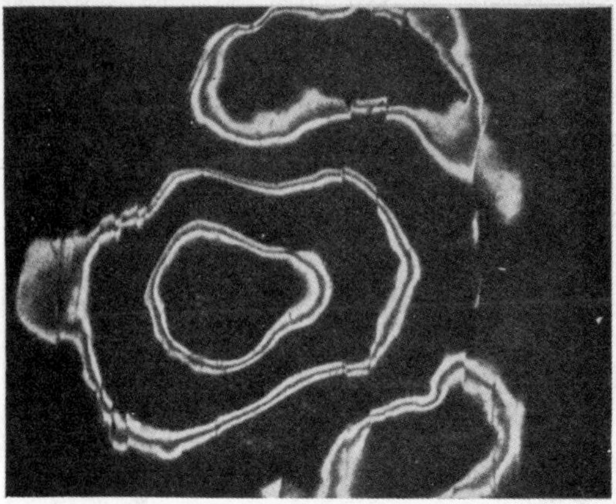

FIG. 69.

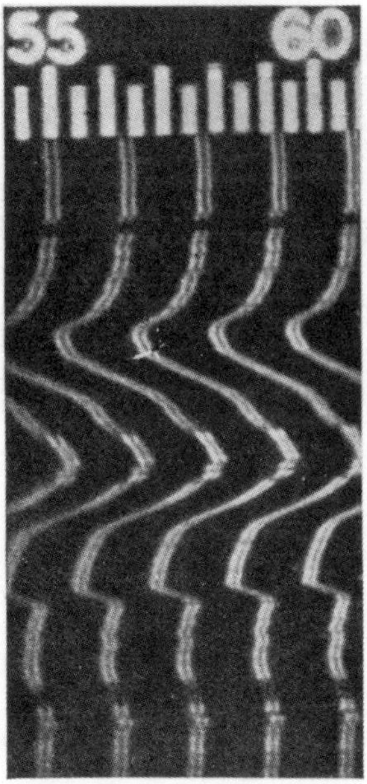

FIG. 70.

Fig. 71 illustrates a further example of an inclusion. The small feature described is shown with two magnifications, the higher of which is $\times 80$. This is probably a growth nucleus and represents a stepped pyramid with the steps of molecular dimensions. Fringes of equal chromatic order across the central circular patch are shown in Fig. 72.

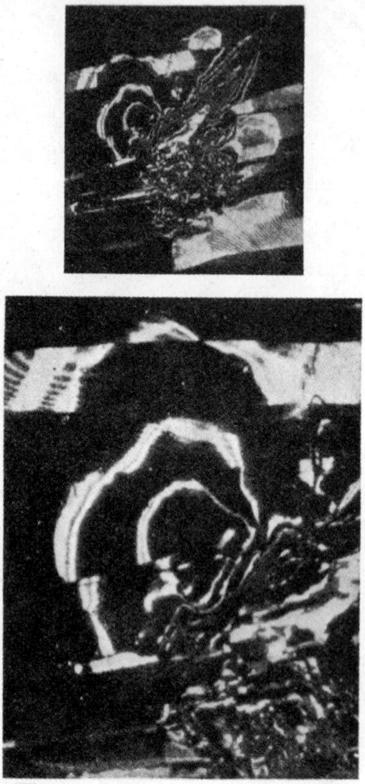

Fig. 71.

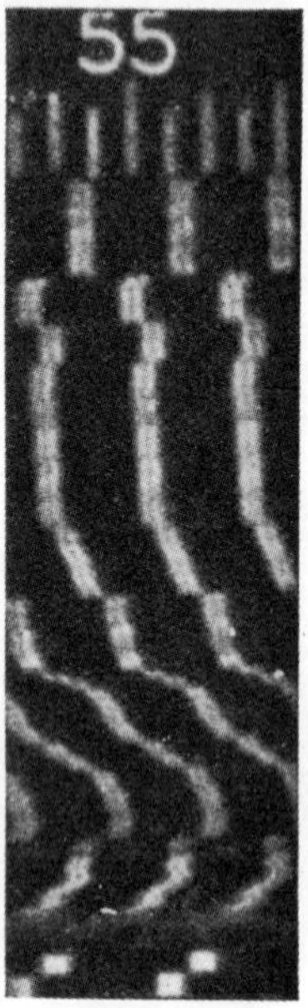

Fig. 72.

In the section following details are given of the application of the methods outlined here to the study of the optical and electrical properties of thin mica sheets, and it is indicated that such properties can be correlated.

CHAPTER X

STUDY OF A RANGE OF MICA SAMPLES

Introduction

MICA is a mineral of considerable industrial importance, particularly as an electrical insulator. Several varieties, muscovite, phlogopite, lepidolite, and biotite, are used commercially and are obtained from widely different geographical regions. For electrical purposes mica is graded visually according to colour and inclusions, the following code being in use:

C.+S.S. . . .	Clear and slightly stained
F.S. . . .	Fair stained
G.S. . . .	Good stained
S. . . .	Stained
H.S. . . .	Heavily stained
D.S. . . .	Dense stained
Spotted. . .	Spotted
A.Q. . . .	First quality
B.Q. . . .	Second quality.

This means of grading does not permit recognition of clear transparent inclusions, and since structural characteristics not normally visible are revealed by multiple-beam interferograms of doubly silvered mica slips, it seemed that use of this technique might afford a valuable supplementary method for the precision grading of mica for electrical purposes. With this in mind, an investigation has been undertaken by P. G. Morris and K. W. Plessner to see whether any correlation exists between the structure as revealed by an interferogram and the electrical power factor and electric strength, covering twenty-five varieties. The outcome of this work is not yet available, and reference will be made here only to a number of structural features which have emerged. Both Fizeau fringes and fringes of equal chromatic order have been employed.

It was first found that the interferogram structure of a slip taken from a given variety is characteristic of that variety within a moderate range of variation. This was done by taking a random dozen slips of mica, of the order of 1/100 mm. thick, from each of several different varieties, and comparing the Fizeau fringe pictures given by the doubly silvered slips. The similarity of the patterns shown by a given variety was apparent. It can be considered, therefore, in the illustrations given here, that any one is typical for its particular mica variety within a moderate range of variation.

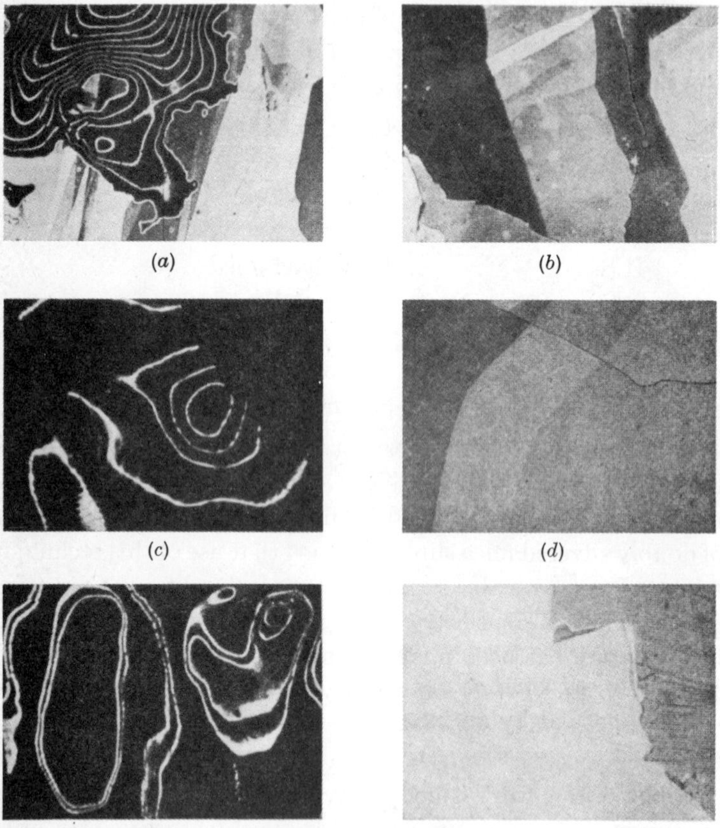

(a) (b) (c) (d) (e) (f)

FIG. 73.

The interferograms

Figs. 73, 74, 75 show interferograms given by a selected number of different varieties. Fuller details have been published in the *Mineralogical Magazine*, and a brief indication only need be given here. The specimens shown in Figs. 73, 74 are all different, e.g. 73 (*c*) is Bihar muscovite, 73 (*d*) is Brazil muscovite, 73 (*a*) is dark amber Travancore phlogopite, 74 (*a*) is a phlogopite, 74 (*d*) a biotite, etc.

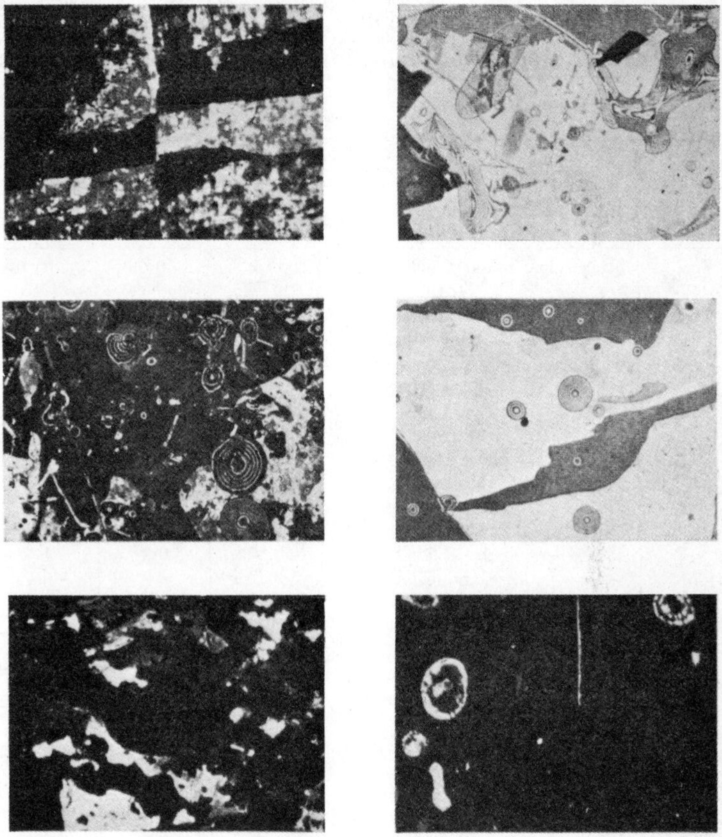

(*e*) (*f*)

FIG. 74.

The distinctiveness of the patterns is striking. In some observations the mica sheet was held between pieces of glass since a sheet 1/100 mm. thick is flimsy, and as a result the uniform tint areas are crossed by secondary low-visibility fringes. It can be recognized that there are three quite distinct features, namely, (1) the variation in average size of the areas of uniform tint, (2) the number of inclusions per cubic centimetre, (3) the average size of the inclusions.

The inclusions are frequently approximately circular, and it is probable that these are liquid bubbles, or originally were, and are now vapour vacuoles. This being the case, it might be expected that a correlation will exist between the number and size of these (often transparent) inclusions and dielectric power loss.

Number	*Designation and origin*		*Inclusions per mm.*^{3}	*Average area of uniformity cm.*^{2}	*Largest uniform area cm.*^{2}
1	Muscovite,	clear green—Madras	17	0·06	1
2	,,	clear green—East Africa	..	0·9	14
3	,,	dark ruby—East Africa	..	0·7	6
4	,,	ruby A.Q.—Madras	..	0·6	5
5	,,	clear green—Madras	..	0·7	6
6	,,	ruby B.Q.—Madras	20	0·1	1·5
7	,,	ruby C.S.S.—Rhodesia	..	0·3	1
8	,,	C.S.S.—Brazil	..	0·9	13
9	,,	C.S.S.—Tanganyika	1	0·5	10
10	,,	C.S.S.—Calcutta	2	0·7	6
11	,,	F.S.—Rhodesia	..	1·0	20
12	,,	F.S.—Brazil	1	0·5	5
13	,,	F.S.—Tanganyika	..	1·3	26
14	,,	F.S.—Bihar	..	1·4	11
15	,,	G.S.—Calcutta	11	0·7	5
16	,,	S.—Brazil	4	0·5	4
17	,,	S.—Calcutta	7	0·7	8
18	,,	Brazil	3	0·4	3
19	Phlogopite,	amber—Madagascar	15	0·06	0·5
20	,,	,,	200	0·01	0·5
21	,,	dark amber—Travancore	52	0·06	0·5
22	,,	Ontario	200	..	..
23	Biotite		55	0·11	0·25
24	,,		100	0·1	0·04
25	Lepidolite		..	0·45	1

Ten samples (all about the same thickness) for each variety were examined and the following determined: (1) average size of uniform tint area, (2) largest uniform tint area, (3) number of inclusions per cubic millimetre. The results given on the opposite page were obtained.

The wide variety in character is of some considerable interest particularly in view of the fact that the eighteen muscovites in this table had all been classified industrially as high-grade transparent micas. They were graded commercially in the regular sequence 1, 2, 3, 4,..., 18. However, it is quite clear that this grading is inadequate. They can be graded more precisely either according to the number of inclusions or according to the average area of uniformity.

Uniform tint areas

A matter of some interest has emerged from the examination of this wide range of micas. It has been found that a few varieties consistently cleave to give relatively very large uniform tint areas. Now it can be shown that uniform tint implies that *the mica has cleaved true to a single molecular plane*. The change in tint produced by a single molecular step can be calculated, in the manner already used in considering the sensitivity of the crossed-fringe technique, thus:

Suppose maximum sensitivity is attained, i.e. let a uniform tint area have intensity $I_{max}/2$, then as before $I = 2/F^{\frac{1}{2}}$ for this region. Now let there be an adjacent step, changing by one molecular lattice space (20 A.U. = $\lambda/273$). The new value δ' leads to a new intensity which can be written I_{max}/K. It follows that

$$\frac{I_{max}}{K} = \frac{I_{max}}{1+F\sin^2\frac{1}{2}\delta'}.$$

From which, using the previous procedure,

$$\delta'-\delta = \frac{2}{F^{\frac{1}{2}}}\{\sqrt{(K-1)}-1\},$$

giving

$$t'-t = \frac{\lambda}{2\pi F^{\frac{1}{2}}}\{\sqrt{(K-1)}-1\}.$$

But for a single molecular step we have

$$t'-t = \frac{\lambda}{273}.$$

Also $$F = 1{,}044.$$

Substitution leads to $K = 4$ (very closely).

Hence the change in height of 20 A.U. reduces the intensity from $\frac{1}{2}I_{\max}$ to $\frac{1}{4}I_{\max}$, i.e. a change of 50 per cent. in intensity is produced, which is a striking result.

When mixed radiations are employed, then many of the uniform areas are suitably illuminated for approximately maximum sensitivity. (If not, the angle of incidence can be varied.) It follows, therefore, that many of the large areas of pure uniform tint are regions which have cleaved *true to a single molecule on both sides.* It view of the fact that areas of over 20 sq. cm. have been found, this is surprising. The crystal must be perfect throughout over the whole of this region. Apart from the crystallographic interest, such mica slips are clearly of value for use as windows or absorbers in alpha-particle and radioactive experiments.

In order of merit, the four muscovite varieties found to be the best from this point of view are given below:

Number	*Origin*	*Average uniform area, cm.*^{2}	*Largest uniform area, cm.*^{2}
13	F.S.—Tanganyika	1·3	26
11	F.S.—Rhodesia	1·0	20
2	Clear green—East Africa	0·9	14
8	C.S.S.—Brazil	0·9	13

It will again be noticed that the commercial grading (the number) does not correlate with the excellence of the material in cleavage.

Examples of uniform cleavage areas (about 1 sq. cm. each) are shown in Fig. 75.

A possible practical application of these observations to interferometry is being developed. Since these surfaces have cleaved true to a molecule, they can be regarded as optically perfect and can be used as the perfect reference plane for interferometric studies of fine detail. When the general topography of a surface, such as diamond, is studied with an optical flat, fine-scale detail can be seen in the fringes and this may be due either to the crystal or the flat. It is intended to employ a selected piece of mica as the reference surface. This will not be flat, and will

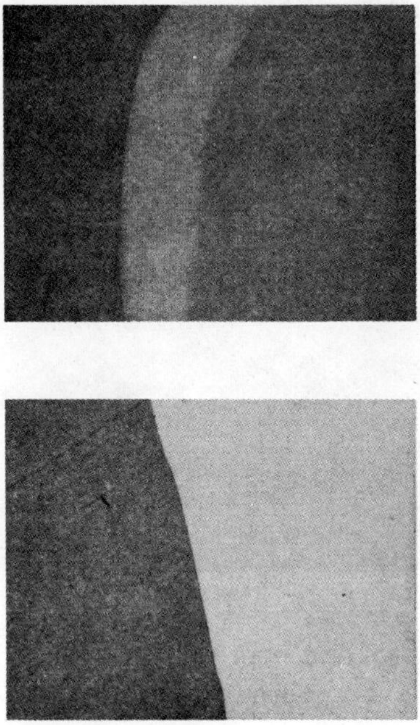

FIG. 75.

produce distorted fringes, but the mica will be free from fine structure and any residual detail must be characteristic of the crystal itself.

The birefringence of the micas

The examination of the micas with fringes of equal chromatic order affords a rapid method for the determination of the birefringence perpendicular to the cleavage plane. The biaxial nature of the mica leads to the doubling of the fringes, the components being plane polarized mutually perpendicularly. As the doubling of fringes does not depend on the thickness, this makes the determination at any given wave-length (of known μ) a simple matter, for any selected piece of mica thin enough to prevent confusion due to overlapping of orders will suffice.

Fig. 76 shows fringes given by a piece of high-grade muscovite, whilst Fig. 77 shows those given by a biotite, and it is notable

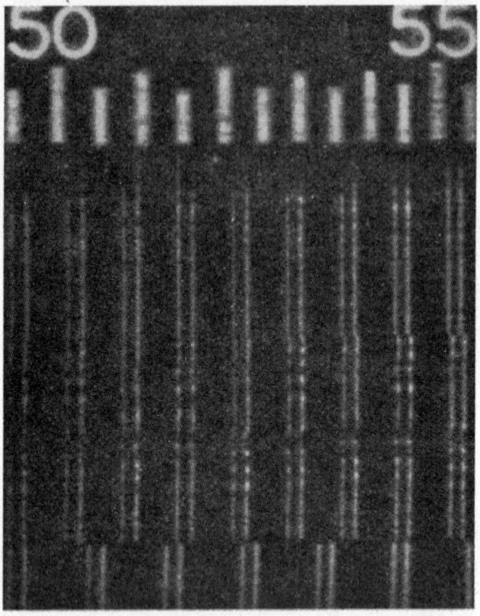

FIG. 76.

that there is no doubling, only a broadening. The former is thus biaxial and the latter uniaxial.

A further point of interest lies in the correlation between the linearity and continuity of the fringes and the mica classification into areas. This is self-evident from the interferograms already reproduced.

Only a brief indication has been given here of the many features observed, particularly with respect to inclusions. Crystal inclusions and even crystal deposits within mother liquors have been recorded. Fig. 78 reveals an included crystal, perhaps haematite, observed in a thin sheet. The rhombohedral plate-shaped inclusion is clearly recognizable. In other samples long chain-filaments of bubbles have been detected, and a complex variety of different kinds of included solid and liquid features have been observed.

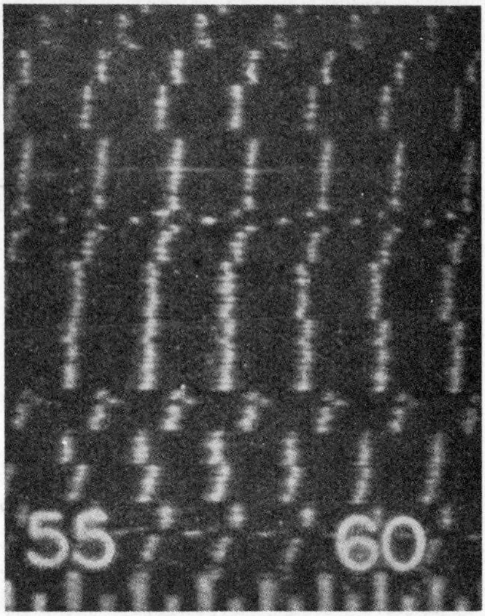

FIG. 77.

FIG. 78.

It is clear that the interferometric method reveals a great deal of unsuspected included matter, and should therefore assist in a more comprehensive study of the optical properties of the micas.

CHAPTER XI

FURTHER APPLICATIONS

In the section following a few typical observations made with various crystals will be selected as illustrating results obtainable by the methods already described. The discussions are not meant to be comprehensive, the object being to illustrate the range of applicability rather than to review information already obtained.

Calcite

Fig. 79 is an example of high-definition fringes of equal chromatic order obtained with calcite. The silvering is thick

Fig. 79.

and the order of interference for the red fringe is 6, i.e. the surfaces are only three light-waves apart. Although dispersion is fairly high, the fringes being separated by almost 500 A.U., yet the sharpness is striking. From the fringe width it is estimated that the reflecting coefficient approaches 0·94. The linear magnification is $\times 100$, and it is considered that inevitable defects on the flat are responsible for the slight wriggling of the fringes. Noteworthy is the small cleaved-out section in the centre. The steps on either side of this are respectively 60 and 72 A.U. Steps of 12, 12, 13 A.U. (± 3 A.U.) have been measured on the same sample in other regions. It will be noted that these steps are all small integral multiples of 12 A.U. The X-ray value for the cleavage spacing between rhombohedral faces of calcite is 6 A.U., from which it is apparent that, as in mica, *the cleavage steps are simple integral multiples of the crystal lattice spacing*. It is not possible to measure a step of only 6 A.U., for this is below the limit of resolution. Yet Fig. 79 shows that the observed step of 60 A.U. which represents only 10 molecular lattice spacings is much in excess of what can be measured by this technique. So far, although microphotometer methods have not yet been applied to this type of fringe, it seems likely that with such procedures a step of only 6 A.U. might be measured.

Selenite

It has been shown in Fig. 37 that the Fizeau fringes of the cleavage face of a selenite crystal exhibit what appears to be some form of 'mosaic' fine structure. This leads to ragged fringes and the accuracy in measurement suffers accordingly. The nature of the mosaic structure is very clearly revealed by the use of the fringes of equal chromatic order. Fig. 80 shows such fringes with $\times 400$ linear magnification (A) and $\times 100$ magnification (B). The surface structure is found to consist of a mass of short, quite narrow strips, generally oriented more or less parallel to the major long cleavage lines.

These narrow strips are not coplanar but vary slightly in height (depth) from 15 A.U. to some 800 A.U. on the particular samples examined. The strip width is usually of the order of 1/10th of the length. Since some of the observed strip lengths are in the region $0{\cdot}5\text{–}1{\cdot}5 \times 10^{-2}$ cm., it seems probable that the microscope has failed to resolve many of the narrower strip widths.

Either fringe of the pair in each picture can be regarded as a contour of the crystal over the region selected by the spectrograph slit. The separation between a fringe pair, corresponding to one order, equals 2,700 A.U., and the various steps in a single fringe are encompassed within a height of only 400 A.U. The vertical traverse across the picture represents 1/5 mm. along the crystal surface. Attention is drawn to the high quality of the definition, the fringe width being quite a small fraction of the order separation.

It is clear that this particular section of the surface consists of a series of alternating up-and-down steps, often of approximately the same heights. The individual short-length features represent strip widths and vary from 0·03 to 0·005 mm.

A number of these 'fine structure' cleavage steps have been

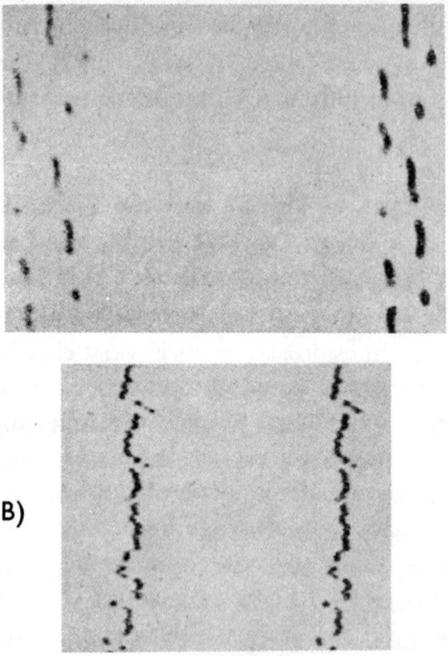

(A)

(B)

FIG. 80.

measured (error some 3 A.U.) and a sample of fourteen of the smaller recorded can be grouped as follows (in A.U.):

	16	28	48	58	77
	14	29	43	..	73
	17	28	..	..	..
	17	31	..	..	..
	14	..	..	..	..
Mean A.U.	15½	29	45½	58	75

The lattice spacing between planes normal to the direction of cleavage as given by X-ray measurements is 15 A.U. It is evident that the above mean values are (within the experimental error) respectively 1, 2, 3, 4, 5, × (15 A.U.). Thus it is established that the fine-structure features are strips stepped by only a few integral multiples of the unit molecular layer, at times, in fact, by but a *single lattice layer*.

This seems to afford strong evidence for the existence of some form of mosaic or lineage structure in the selenite. If it can be supposed that cleavage is true to a molecular plane within a perfect crystal, then each elementary strip can be considered as a perfect crystal. The cleavage, being true to a molecular plane in such a unit, jumps at the boundary to form a step. The units vary considerably in area, from larger than $0{\cdot}03 \times 0{\cdot}3$ mm. to below $0{\cdot}005 \times 0{\cdot}05$ mm. It is not possible to give any indication of the heights of the units since cleavage can occur at an arbitrary level within a unit and the cleavage steps need not necessarily be the heights of the blocks. Attention may be drawn to the fact that, as in mica (despite the smaller lattice spacing), once again it is possible to evaluate a crystal lattice spacing with visible light-waves by virtue of multiple-beam interferometry.

Diamonds

It has been shown in Chapter VI that the crossed Fizeau fringe technique has revealed considerable detail about the topographical structure of the surface of a natural diamond face. An advantageous combination can be adopted by using crossed Fizeau fringes to obtain a general survey of a crystal face and then from such pictures individual details can be selected for more elaborate examination by the more convenient fringes of equal chromatic order. The advantages of the latter are evident, for the natural rapid variations in dispersion locally inherent in a complex Fizeau topography are avoided. This rapid variation is the principal factor responsible for inaccuracies in calculation, as has been stressed before. The fringes of equal chromatic order do not suffer from this defect. Furthermore, the dispersion of the fringes is under control, whilst the dispersion of the Fizeau fringes depends upon local curvature and frequent situations arise in which the dispersion cannot be increased beyond a specific limit.

An excellent example of the application of this combined technique is shown by the examination of the triangular feature shown in Fig. 81. This feature is fairly small, for the triangular marking is only 1/5 mm. across. The fringes of equal chromatic order for a vertical line section over the ridge and through the

Fig. 81.

triangle are shown in Fig. 82 (but with lower magnification). A number of interesting features may be noted. A slight convex curvature can be seen in the base of the triangle (the fringe second from the bottom). Furthermore, the surface area leading to the triangular pit dips down, though the dip is very slight. In order to show how powerful the method can be a very high-dispersion picture of a section across the triangle is shown in

Fig. 83. The fine-grain details visible in both Figs. 82 and 83 can be shown to be due entirely to the optical flat used as the matching surface. Fig. 84 represents crossed Fizeau fringes from another diamond surface, and it will be noticed that the

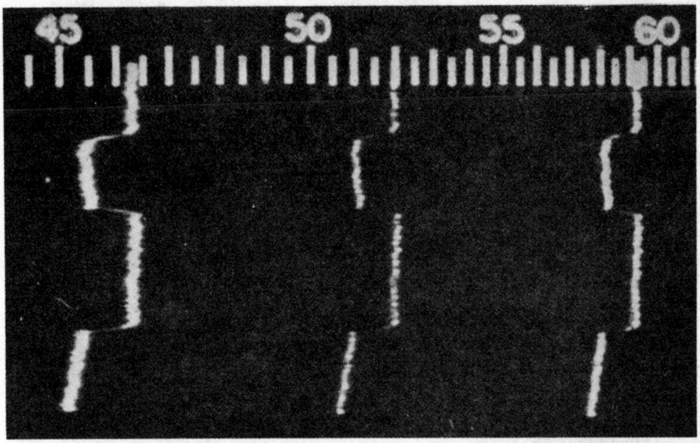

FIG. 82.

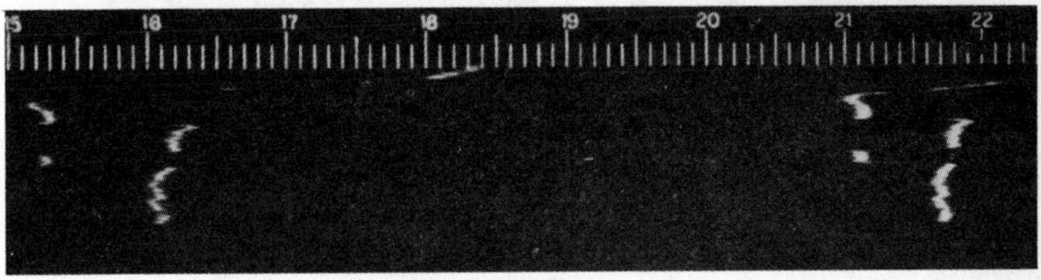

FIG. 83.

characteristic triangular markings are accompanied by projections on the base. Fig. 85 shows the fringes of equal chromatic order crossing a triangle on a line parallel to one of these base projections. This reveals the slight drop on the surface as the triangular pit is approached, a feature not so readily realized from the crossed Fizeau fringes.

Many further beautiful examples of the combined examination by crossed fringes used in conjunction with fringes of equal chromatic order have been obtained and features of considerable crystallographic interest revealed.

FIG. 84.

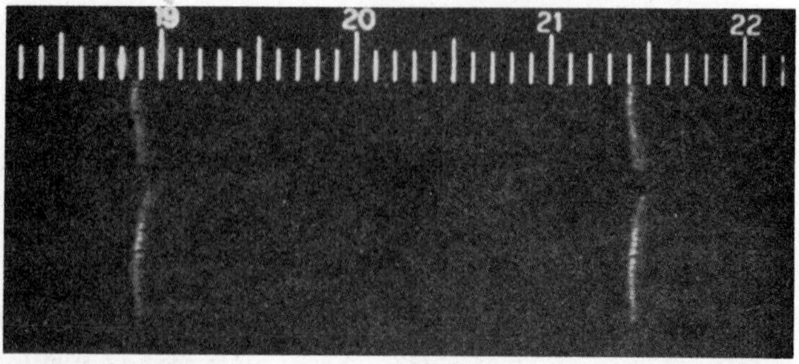

FIG. 85.

CHAPTER XII

OPAQUE SUBSTANCES AND METALS

General

In all the previous cases treated it has been implicitly assumed that the object studied is transparent, thus permitting the observation of fringes in *transmission*. There are, however, very many opaque crystals, and, furthermore, the surface phenomena of polished and etched metal surfaces are of considerable interest

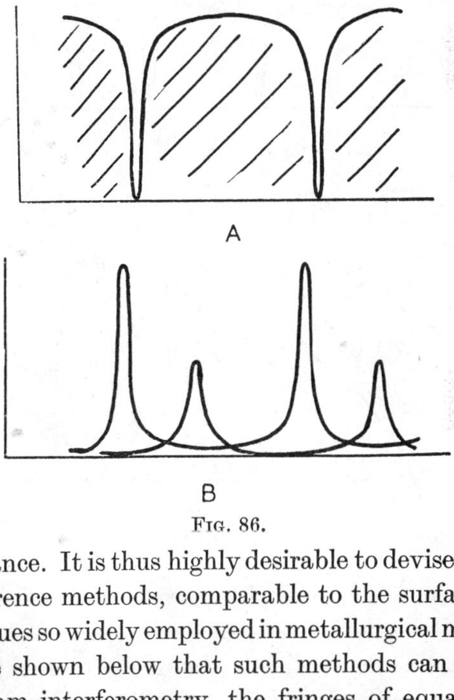

Fig. 86.

and importance. It is thus highly desirable to devise back-reflection interference methods, comparable to the surface-illumination techniques so widely employed in metallurgical micrography.

It will be shown below that such methods can be used for multiple-beam interferometry, the fringes of equal chromatic order being the easier to apply. The nature of the back-reflected multiple-beam Fizeau fringes will first be considered. It can readily be demonstrated from the treatment in Chapter II that the multiple-beam-transmitted Fizeau fringes are accompanied by a complementary reflected Fizeau system consisting of narrow dark fringes on a broad bright background, resembling absorption lines, as is shown in Fig. 86 A. If reflected light instead of transmitted light is employed, sharp fringes are still obtainable, but it is necessary to monochromatize the light. Suppose the light contains two wave-lengths, then in transmission these appear separated as in Fig. 86 B. In reflection, the result is as shown in Fig. 87 A, from which it is seen that the broad background light continuum associated with each wave-

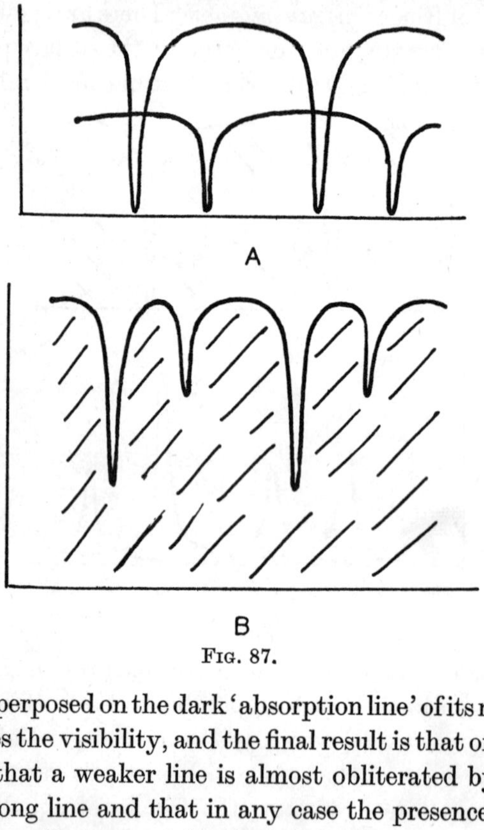

FIG. 87.

length is superposed on the dark 'absorption line' of its neighbour. This reduces the visibility, and the final result is that of Fig. 87 B. It is clear that a weaker line is almost obliterated by a neighbouring strong line and that in any case the presence of neighbouring lines affects the definition of the fringes.

This restriction to monochromatization reduces the flexibility somewhat since it is not possible easily to isolate the yellow mercury lines free from the green. The green line can be isolated with a filter, but in order to avoid errors due to overlap of orders it is necessary to employ more than one wave-length. When this is done the fringes suffer in consequence. This must be tolerated, and then the precision measurements are made with the isolated green line, once order allocation has been established. Also the valuable crossed fringe technique cannot be adopted.

Theory of reflected fringes

The Airy sum for the reflected system is illustrated in Fig. 88, in which R and T are the fractions reflected and transmitted at the films. There is one intense beam, the first, with intensity

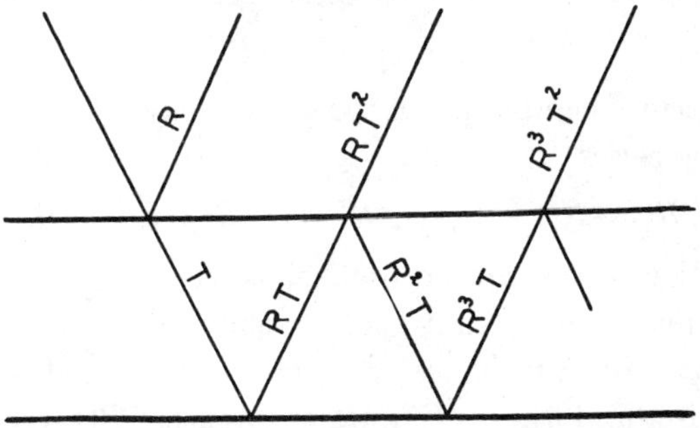

FIG. 88.

of the order of 90 per cent., and this is followed by a series with amplitudes $RT^2(1+R^2+R^4+...)$. If there is no absorption, T is 0·1, and thus it is clear that the first beam predominates. If the full Airy summation is carried through the fringe shape is that of Fig. 86 A, obeying an exactly complementary relation to the transmitted system. *The minima go down to zero* and the broad maxima have the value

$$I_{\max} = \frac{4R}{(1+R)^2},$$

which is very close to unity when $R \geqslant 0{\cdot}90$. Thus for $R = 0{\cdot}90$, $I_{\max} = 0{\cdot}998$, and even for the low value $R = 0{\cdot}75$, $I_{\max} = 0{\cdot}98$.

It is to be noted that the classical Airy sum disregards any anomalous phase-change effects which may arise from the fact that the first intense beam does not go *through* any silver film whilst all the remaining beams pass through two film thicknesses. It is important to note that *when there is no absorption, the fringe minimum is always zero, no matter what the reflecting coefficient.*

Effect of absorption

The above conclusions are in fact far from what is observed, and the cause of the discrepancy is the neglect of the effect of absorption. Since this has not been discussed in the literature on interferometry it will be briefly reviewed here. Without absorption $T+R = 1$, but if a fraction A is absorbed in passing through a single film, $T+R+A = 1$, hence the transmission is reduced from T to $T' = KT$, in which K is less than unity and equals $\{1-A/(1-R)\}$. The reflected light can now be written

$$R+K^2RT^2(1+R^2+R^4+...).$$

The first term R can be split into two, namely, $R(1-K^2)+RK^2$, making the series

$$R(1-K^2)+K^2[R+RT^2(1+R^2+R^4+...)].$$

The quantity in the square bracket is the Airy sum. Clearly, then, the net result is that there exists a classical Airy fringe system, *but reduced in intensity by the factor* K^2, and superposed upon this is a uniform continuum of intensity $R(1-K^2)$. [Note that if A is zero $K = 1$ and the continuum disappears, leaving the classical distribution.]

Now A, and therefore K, depends upon the reflectivity in the manner discussed in Chapter III. As R increases K diminishes, since A and $(1-R)$ approach each other as T gets smaller. The influence of the two separate terms discussed above is shown by Fig. 89. Here, on the same scale, is a plot of the way in which the fringes and the superposed continuum vary with the reflection coefficient. The fringes have an intensity maximum $K^2 4R/(1+R)^2$ which closely enough equals K^2, and the variation of this with R (due to the increase in A with film thickness) is shown in curve I. This treatment is a rough approximation, since amplitudes and not intensities should be combined.

Below this, curve II shows the way $R(1-K^2)$ increases with R. The sum of the curves I and II is the straight line III which represents the total reflected light in the broad fringe maxima. This equals $R+K^2(1-R)$ and is nearly constant. The amount of light in the reflected maxima is thus practically unaffected by the absorption, but the minima do not go down to zero.

The amount of unwanted light reducing the fringe visibility is shown by the shaded area below curve II. The ratio of the ordinates gives a measure of the dilution of the fringes by the unwanted light. Thus at reflectivity of 85 per cent.

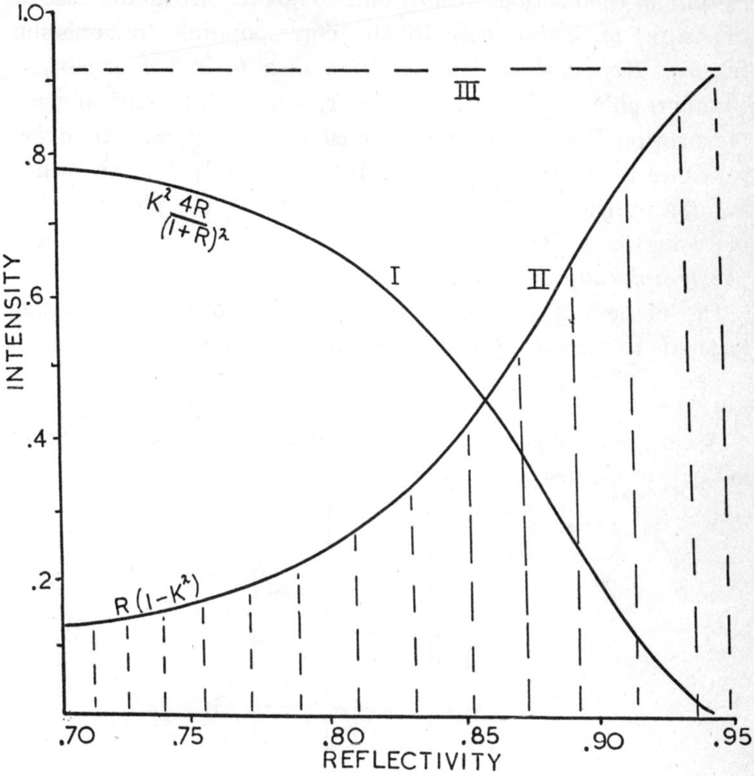

FIG. 89.

this ratio is unity. At reflectivity 90 per cent. it is 3:1, an amount that can still be tolerated. At reflectivity 94 per cent. the ratio is 50:1, as a result of which the fringes are hardly visible. These estimates are rough approximations.

From the curves it appears that a reflection coefficient up to 0·85 on the upper surface can be tolerated. This leads to narrow fringes, particularly because a higher value of reflection coefficient can be adopted for the lower opaque surface, say, 0·95.

Since fringe width is determined by $\sqrt{(R_1 R_2)}$, where R_1 and R_2 are the respective reflectivities of the two surfaces, the net effect is that an *effective* coefficient of 0·90 can be used.

It is particularly to be noticed that these high reflectivities, in spite of their serious absorption, do not require an increase in exposure, as is the case in the corresponding transmission fringes. *Exposures are therefore very short for reflection fringes.* Photographic exposure is, however, much more critical with 'absorption lines' than with transmission fringes. An over-exposure tends to obliterate the fringes entirely. On the other hand, a judicious exposure (by making use of the photographic plate lag) can give a picture in which the fringe minima are *photographically* close to zero.

In succeeding sections the application of back-reflection methods to various special cases will be considered.

Quartz

A simple arrangement for use either with a camera lens or low-power microscope is shown in Fig. 90. A parallel beam from

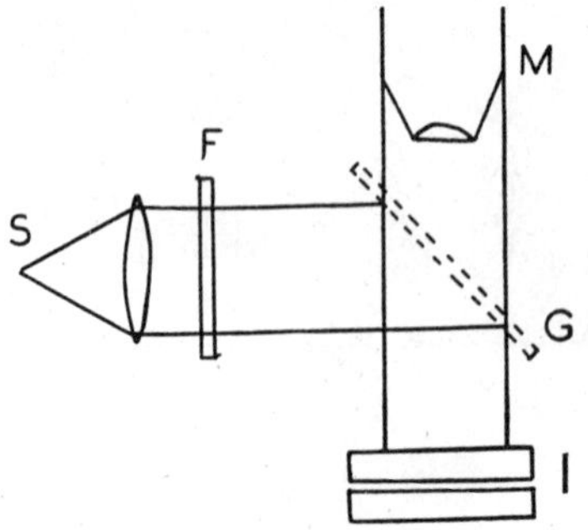

FIG. 90.

the mercury source S is filtered by F and then reflected by the glass plate G on to the interference system I. The reflected fringes are photographed by the microscope M which is focused on I. The procedure is only applicable with low-power objectives giving a sufficient working distance MI to permit introduction of the plate G.

The result of using this method on the quartz crystal for which transmission fringes were described in Chapter V is shown in Figs. 91, 92. There is complete identification between the transmitted and reflected patterns. The definition is slightly superior in the reflected system because the collimation is more critically

FIG. 91.

correct. For in transmission, the light must pass through the crystal, entering at an irregular surface. Any surface and internal defects contribute to reducing definition. Further, the silvering on the lower face (the face to be studied) can be made thicker than is possible with transmission fringes. A slight overexposure tends to 'cut' the wings of the 'absorption line' and thus produces spurious sharpening. This is a somewhat dangerous procedure, for, strictly speaking, the fringes are slightly asymmetrical and such a spurious sharpening might be associated with a slight displacement of the effective optical 'centre of gravity'. Even if it exists, however, this effect will probably take place equally in similar fringes and it is not likely to be serious.

A simple direct method for examining reflection fringes with pure monochromatic sources is to make use of a spectroscope as shown in Fig. 93. The source A is a mercury arc, and either

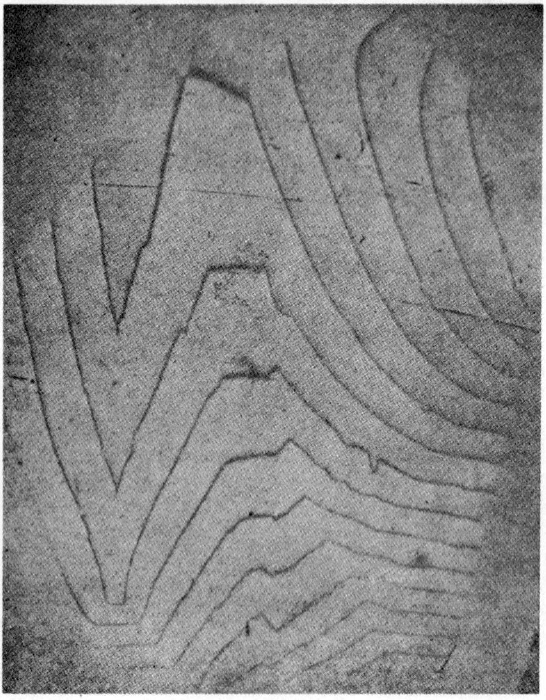

FIG. 92.

a very wide slit can be used or the slit can be removed entirely.

One sees, then, fringes from the surface XY in the spectrograph at PQ, somewhat as shown in Fig. 93 (A). L, M, N, O are the patterns of the yellow doublet (LM), the green line N, and the blue line O. An advantage of this system is that an alternative white-light source at B permits fringes of equal chromatic order to be observed by a quick change-over of the reflector between lens and interferometer, followed by suitable adjustment of the lens L.

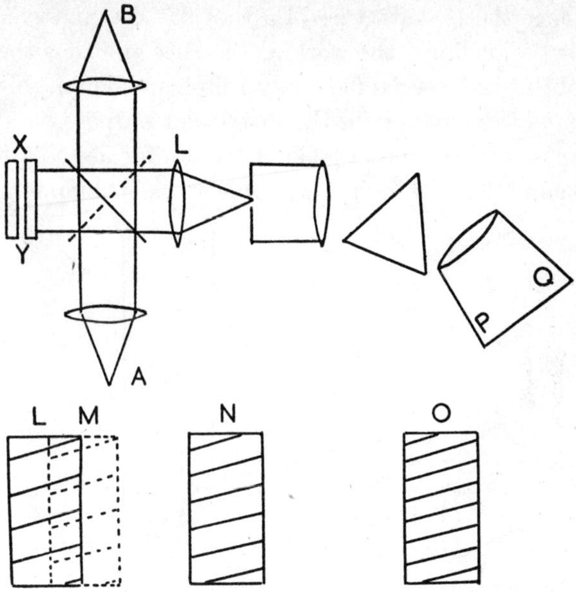

(A)

FIG. 93.

Microscope technique

The examination of the reflected fringes with a medium-power or high-power microscope leads to difficulties. The lens working-distance being small, it is necessary to employ surface illumination by one of the methods used in metallurgy, with the restrictions imposed by precise collimation. One satisfactory method is to use a glass reflector as in Fig. 94 above the microscope objective.

For collimation an image of the source I (filtered by F) must be formed at the rear focus of the objective O. By this means O directs a parallel beam at normal incidence on to the interference system X. However, O must simultaneously form an image of X. A normal microscope objective is not designed to do both functions simultaneously and thus the collimation suffers; in any case adjustment of I at the rear focus is a matter of difficulty. A further difficulty is imposed by the need to use an optical flat above the metal surface. Flats are usually fairly thick, since the manufacture of a thin flat is a matter of some difficulty. This limits the working distance and thus the power of the objective. Special flats only 1 mm. and 3 mm. thick have been found satisfactory for this particular purpose.

It is possible to select a good cover-slip for use as a flat and thus to employ fairly high-power objectives. Oil immersion has

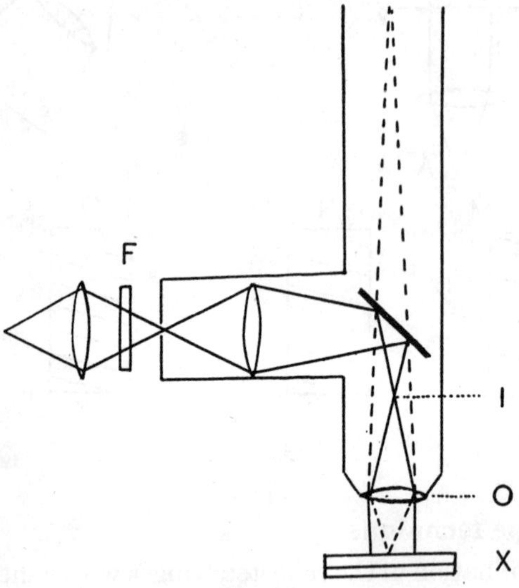

FIG. 94.

not yet been tried, but there appears to be no objection to it. An alternative under consideration is the use of a thin selected mica slip as cover-glass 'flat' if fine structure is to be investigated. With such means high powers can be used. It has already been pointed out that very high powers cannot be employed with Fizeau fringes. High powers can, however, be used with back-reflected fringes of equal chromatic order.

Fringes of equal chromatic order with metals

The production of fringes of equal chromatic order in back reflection is relatively easy if low powers are needed, and the problem of monochromatizing met with in Fizeau fringes does not exist. The identical system as with Fizeau fringes (but with different source and no filter) is used, the image of the inter-

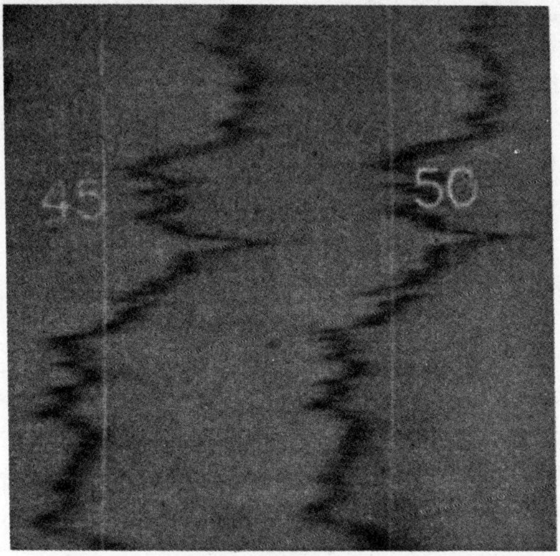

FIG. 95.

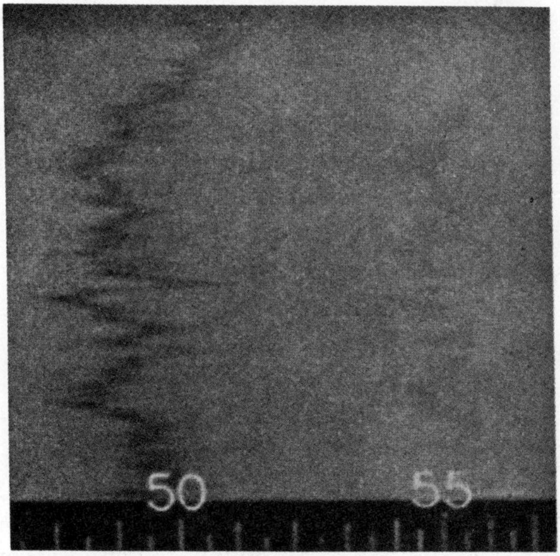

FIG. 96.

ference system being projected on to the spectroscope slit in both cases.

Experiments in this field have only just begun, yet the results so far obtained justify the optimism that the technique has a useful contribution to make to many of the problems connected with metal surfaces.

As an example Fig. 95 shows the fringes observed with an *unsilvered* highly polished stainless steel slip gauge. The linear magnification is $\times 300$. The fringes represent a contour which in this photograph is some $\times 300,000$ in the vertical direction. This can be compared with electron-microscope studies which give the main magnification along the surface, not normal to it as is the case here.

The detail in this photograph is such that useful information about surface structure may be made available if correctly interpreted. Fig. 96 shows another section of the same surface, and it is evident that the fringes are revealing the quality of the surface polish.

Phase effects

A fundamental distinguishing feature in the case of metallic surfaces is the question of anomalous reflection phase effect. A polished steel surface or an etched surface may present areas metallographically distinct. It is perhaps possible, though not yet established, that reflected light even at normal incidence may exhibit differential phase change. When light is reflected from a metal at normal incidence there is a change in phase, perhaps of the order of π. If different areas on an otherwise plane surface have different phase changes the interference picture will produce a spurious, apparent *geometrical*, structure if the phase effect is not recognized. For a change in phase leads to a shift in fringes, and this would be wrongly interpreted as a geometrical height alteration.

Whether in fact this happens is now under examination, and two methods have been devised for eliminating (or alternatively demonstrating the existence of) this effect. The first method is to evaporate silver over the metallic surface, as for the general case. This heightens the natural reflecting coefficient and deposits a uniform silver film over the different features and should eliminate the phase effect. The phase effect might only arise in those exceptional cases where the natural polish of the metal is so high that silvering can be avoided. This is the exception, and in general, since silvering will be used, the phase effect may not be serious.

The second method for avoiding phase effects, and indeed for simplifying the examination of metals, is the plastic moulding technique described in the next chapter.

Thickness of a thin metallic film

The thickness of a very thin reflecting metallic film is difficult to measure. Previous determinations of film thicknesses of the

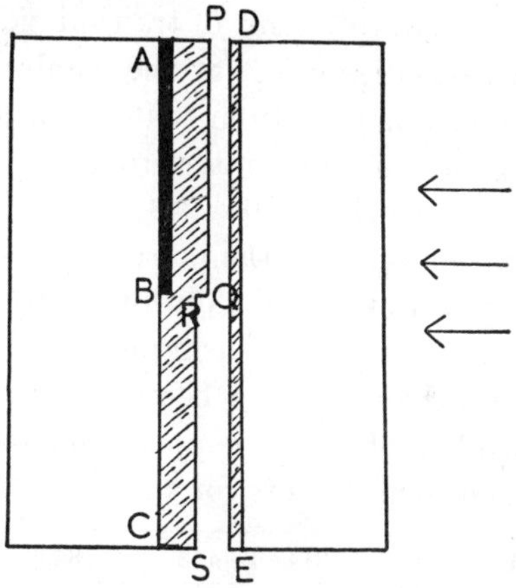

FIG. 97.

order of, say, 1 → 0·01 light-wave have usually been made either by weighing or chemically estimating the quantity of metal on a known area. These methods are not only difficult because of the small quantities involved, they also suffer from the defect that one must assume the density of the tenuous film (of atomic dimensions) to be that of solid metal, an assumption requiring careful checking. Multiple-beam methods developed by W. K. Donaldson and A. Khamsavi have solved the problem in the following manner.

The film *AB* (Fig. 97) is deposited on part of a flat glass surface *ABC*. A fairly thick (opaque) coating of silver *PQRS* is then deposited over the flat. The height *QR* then equals that of the silver film AB. A silvered flat DE ($R = 0{\cdot}90$) is brought close to the combination on AC. Illuminated from the right, back-reflected Fizeau fringes are formed and one obtains a picture like that given by a mica cleavage step. The step displacement RQ can thus be determined with a precision down to the order of 10 A.U.

It should be noted that the *true metrical thickness of the film is obtained.* The coating of the film and remainder of the flat AC with the opaque film $PQRS$ automatically eliminates phase-change effects which would vitiate results if transmission fringes were used. It is not sound practice to make the coating film $PQRS$ sufficiently thin to permit of transmission observations, for then the phase-change effect at the combination PQ may differ in an unknown manner from that at the simple film RS.

Fig. 98 shows the reflection interferograms obtained by A. Khamsavi with film thickness 66, 214, 520 A.U. respectively. These fine sharp fringes reveal the power of this multiple-beam method for the measurement of thin films. By estimating the mass of silver on a known area using a microchemical colorimetric technique, Khamsavi has been able to determine the densities of thin films of silver, copper, and aluminium. The following table shows his results for silver.

Thickness, A.U.	122	314	448	518	520	735
Density	10·4	10·6	10·5	10·2	10·2	10·5

Within the experimental error (0·3) all have the same density as block silver (10·5).

By employing this interference method, but using fringes of equal chromatic order instead, Donaldson has derived an experimental relation between reflectivity and film thickness for silver. The silver films, produced by evaporation, were used as reflectors in a spectroscopic reflectivity measuring instrument and reflectivities, transmissions, and absorptions measured. They were then treated in the manner described, being coated over by silver and the thickness determined. These results have been used in the earlier chapters giving reflectivity data.

It is clear that this thin film technique has distinct possibilities, for it need not be restricted to metals. The film *AB* in Fig. 97 can be of any kind capable of withstanding the vacuum necessary for evaporation of the superposed coating of silver. It will

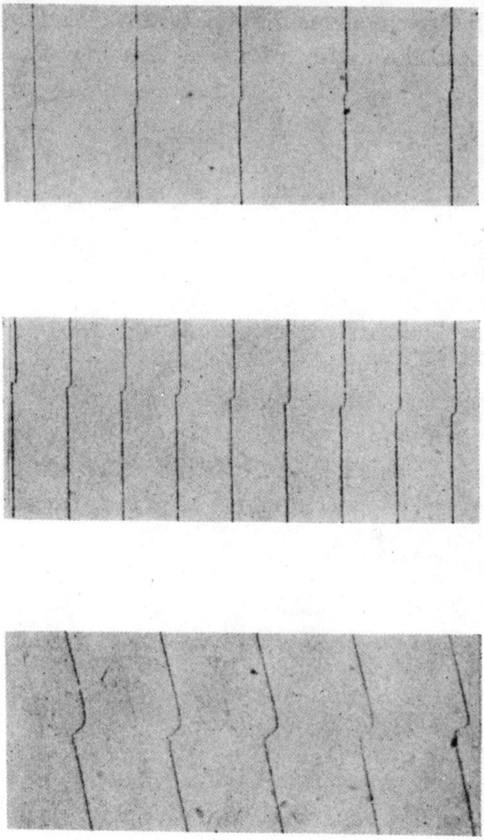

FIG. 98.

be shown in Chapter XIV that cryolite films can be used for a special interferometric purpose and this present technique can be applied to the measurement of the thickness of such films.

Experiments are now in progress with other films, notably protein films of biochemical interest, and from this mode of investigation the following features might emerge: (1) thickness of monolayers, (2) local variations in monolayers, (3) birefringence of monolayers. The method can only be judiciously employed with organic material because the film must be exposed to a vacuum for silver deposition.

Investigations are also possible by enclosing thin films of liquids between silvered flats. Both Fizeau fringes and fringes of equal chromatic order have been observed under such conditions.

CHAPTER XIII

PLASTIC REPLICAS

The need for a replica technique

It has been demonstrated that the examination of an almost flat surface of a transparent body (e.g. a clear crystal or a thin film) by multiple-beam interferometry is relatively simple, and that fine sharp precision fringes are obtained if the two surfaces concerned are suitably silvered and maintained at a separation no greater than a few wave-lengths of visible light. If a study is to be made of an appreciably concave surface, on the other hand, or of features lying below the general level of the specimen (e.g. deep etch pits), then the condition of close approach cannot be realized. If, however, a reliable negative replica cast can be prepared, the originally depressed features can be elevated and brought within the requisite distance from the reference flat.

The application of a reliable replica technique is of greater value in the examination of opaque surfaces, such as those of metals, as the direct interferometric study of a metallic surface is subject to the following drawbacks.

Because of opacity a back-reflection technique must of necessity be employed. If a low-power objective is in use, the working distance is sufficient to allow the insertion of a 45° reflector between the objective and the optical flat and no great difficulties are involved, but the method is only of practical value for useful magnifications not exceeding $\times 50$. If higher magnifications are required and the surface illumination practised in standard metallurgical microscopes is resorted to, the reflector is between the objective and the eyepiece, the former performing the dual role of condensing light on to the specimen and at the same time forming an image of the illuminated surface. However, for multiple-beam interferometry the light falling on the specimen must be fairly strictly parallel and at normal incidence. Hence the microscope objective must simultaneously produce a parallel beam from a small image formed at its rear focus and also act in its normal capacity as an image-forming lens. To carry out both functions correctly it would appear that a specially designed lens is necessary.

It has been shown that the visibility of the reflected fringe system is critically dependent upon the light-absorption of the silver film deposited upon the reference flat. If effective reflectivities as high as those permissible with transmission fringes are employed with the back-reflection technique, the fringe visibility is poor, and consequently the fringes are hard to detect. The visibility can be improved by lowering the reflectivity of the silver film on the flat, but this leads to an increased fringe width.

The fact that the reflected fringes are fine dark lines on a broad bright background necessitates the use of monochromatic radiation, which is a drawback since considerable advantage in interpretation arises from the use of a mixed group of a small number of distinct wave-lengths. This restriction to monochromatization means that a series of separate photographs must often be taken, each with a different wave-length, thus increasing both the labour and the difficulty of interpretation.

A principal difficulty, which can lead to serious errors, arising when metal surfaces, particularly alloys, are under examination, is the question of differential phase change at reflection.

There would appear to be two ways of avoiding this error. One method would be to evaporate a fairly thick film of silver over the metal surface, and thus impose a uniform phase change (that for the silver) over the whole area. This appears attractive, but there may also be an error involved in so far as it is not known whether regional variations beneath the silver (e.g. corrosions) locally affect the phase change. (It is recalled that the silver film will be less than 1,000 A.U. thick.) The second method for overcoming the difficulty is to develop a technique for making a reliable transparent replica casting, which eliminates phase effects.

Thus, if a transparent replica technique can be evolved, the difficulties discussed above will be removed and the simpler methods available for transparent bodies can be applied to the study of the surfaces of opaque solids. The realization of this possibility would be a considerable advantage.

The experimental technique

A replica is required with the following characteristics:

1. It must be transparent.
2. Features in extension (i.e. across the surface), the dimensions of which are at least as small as 1/500 mm., must be reproduced with high fidelity.
3. Features in depth must be reproducible to within approximately molecular dimensions.
4. As the replica is to be handled for adjustment it should be robust.
5. It should be capable of remaining in a vacuum without distorting.
6. A copious gas stream must not be liberated from the replica when in a vacuum, otherwise a silver coating of low light absorption cannot be deposited upon the surface.

Two materials which it was considered might meet these requirements were the I.C.I. plastic products Transpex I (unplasticized polymethyl methacrylate) and Transpex II (unplasticized polystyrene). Of these the methacrylate polymer is to be preferred since it is far more readily degassed in vacuum than the polystyrene, and all the measurements and photographs presented here were made with Transpex I replicas. The material employed is in sheets about 4 mm. thick. From such a sheet a small piece is cut and washed with soap and warm water and then transferred to an oven, where it is maintained at a temperature of about 170° C. At this temperature the plastic rapidly dries and is brought to a state suitable for moulding. Drying with a fabric is not attempted since frictional charges are readily built up, resulting in the collection of fibres and dust. The specimen, mounted in a suitable cement, is placed on an electric hot-plate and brought to a temperature of 140° C. The plastic is placed on the specimen and then covered with a piece of hot plate-glass. Upon the application of slight pressure the soft plastic flows and moulds itself to the specimen surface on the one hand and to the plate-glass surface on the other. The plate-glass is depressed until it finally rests on an accurately machined brass ring supported on a plane containing the specimen face. Consequently the finished replica approximates closely to a plane parallel sheet. This is desirable as the presence of an appreciable wedge angle between the faces of the replica gives rise to ghost images and secondary fringes. The whole is allowed to cool slowly (at least an hour) to room temperature to avoid the setting up of internal strains. The replica and specimen are easily separated, the former then being coated with an evaporated silver film. Due to the flow characteristics of the plastic, the moulding pressure employed is quite small, and it is not considered that specimen deformation is introduced.

The specimen is mounted according to its character in a suitable cement capable of withstanding the moulding temperature. For this purpose an artificial stone, Kaffir 'D', has been found generally useful. An alternative investment is a phenol-formaldehyde thermosetting resin. Prior to the preparation of a replica it is helpful to clean the specimen by employing the well-known method of repeatedly stripping off collodion films.

Reproduction in extension

To test the reliability of the reproduction in extension casts were made from plane gratings ruled on metal. The replica gratings (which are quite robust and far superior to the usual Thorpe replicas found in teaching laboratories) were mounted on a spectroscope and the number of lines per inch determined with the Na D lines. The values so obtained are compared below with the data marked on the original gratings.

Lines per inch

Original . . .	2,400	14,200	17,300
Replica. . . .	2,412	14,270	17,370
Difference, %. . .	+0·5	+0·5	+0·4

Of particular interest is the fact that the 17,300-line metal grating showed symmetrical ghost lines whose intensity increased with order number. Identical ghosts were observed with the replica grating. Furthermore, the definition and resolution of the replicas were in each case identical with those of the parent grating. It is clear that in these three cases the lateral structure of the specimen surface was reproduced by the replica, the whole scale of the structure having undergone, however, a contraction of about $\frac{1}{2}$ per cent.

The $\frac{1}{2}$ per cent. contraction arises from the difference between the thermal expansion of the metal and that of the plastic. The plastic tends to conform to the specimen structure down to the setting-point (about 100° C.), at which state it will possess the structure of the thermally expanded metal at that temperature. As the coefficient of thermal expansion of the metal is about $0{\cdot}2 \times 10^{-4}$ per ° C. over the range 100° to 20° C., it contracts by about 0·2 per cent. The expansion coefficient of the plastic is higher, being about 1×10^{-4} per ° C.; consequently the plastic contraction is some 0·8 per cent. The difference between these two contractions, 0·6 per cent., represents the net shrinkage of the cooled plastic relative to the final cold state of the specimen. This figure is in close enough agreement with the measured values for the effective contraction. It follows that a sufficiently accurate correction can always be made if 1×10^{-4} per ° C. be taken as a mean linear coefficient for the plastic. In the majority of cases this small correction can be disregarded.

Reproduction in depth

Whilst the many well-known replica techniques developed for electron microscopy indicate that faithful reproduction in extension might have been anticipated, such experience offers very little evidence for the expected behaviour of a replica in terms of depth. It is this aspect which is the crucial one for interferometry, as it is in the determination of small heights and depths that the multiple-beam interference technique is so specifically powerful. It is found that the reproduction in depth is very close indeed and sufficient for many purposes. Replicas were made of surfaces the characteristics of which were already known from previous interference experiments, and the replicas then compared with the originals.

1. *Glass*

Fig. 99 shows the multiple-beam Fizeau fringes (λ 5,461) formed when a replica of a piece of thin glass has been silvered and matched against a similarly silvered glass flat. The magnification is $\times 40$. The fringes are typically those shown by glass. It has already been demonstrated (p. 74) that the fine disrupted structure of the fringes is due to polish marks on the reference flat, this structure appearing only under critical illumination conditions. Fig. 100 is a further example, revealing a trapped air-bubble, which appears rarely when making a mould.

2. *Mica*

It has previously been established that the steps appearing on the cleavage faces of muscovite are often small integral multiples of the 20 A.U. lattice spacing. A sample of muscovite was baked at 120° C. to drive off included water, and a replica was then taken from a freshly cleaved face. Both the original and the copy were silvered and examined, using Fizeau fringes. The characteristic cleavage steps were reproduced on the plastic, these steps being compared with the corresponding ones on the mica. Measurements on the replicas showed that in this case also the step heights were true to within 5 A.U., the error in observation. The replica is so exact that it is quite easy to identify cleavage lines, and a precise comparison between the step heights on the mica on the replica can be made. The results given in Table VI have been selected arbitrarily from such measurements, the step heights being in Ångström units. The experimental errors in each of the values quoted is of the order of 5 A.U.

TABLE VI

Mica . . .	385	505	1,026	1,280	1,318	3,237	4,865
Replica . .	385	499	1,020	1,274	1,302	3,196	4,822
Difference, % .	..	−1·2	−0·6	−0·5	−1·2	−1·3	−0·9

It is seen that over an extensive range the replica steps closely follow the original mica ones, but are consistently smaller than the latter by about 1 per cent. This shrinkage is not detected

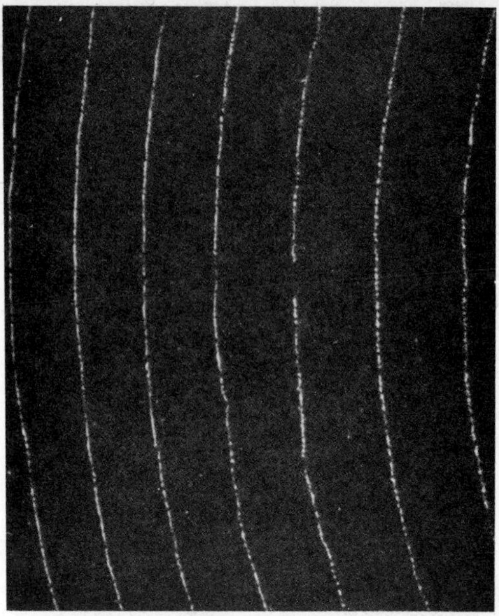

Fig. 99.

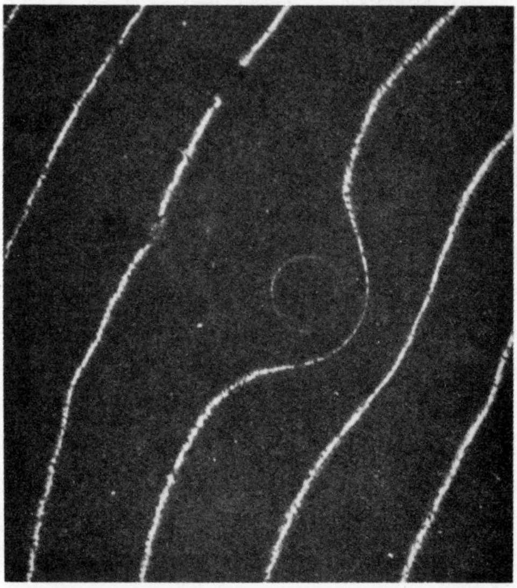

Fig. 100.

in the first step, being masked by the somewhat larger experimental errors. The 1 per cent. shrinkage in depth is consistent with the shrinkage found in extension with the metal gratings, since the expansion coefficient of mica is considerably less than that of a metal.

Figs. 101, 102 show Fizeau fringes given by replicas of mica.

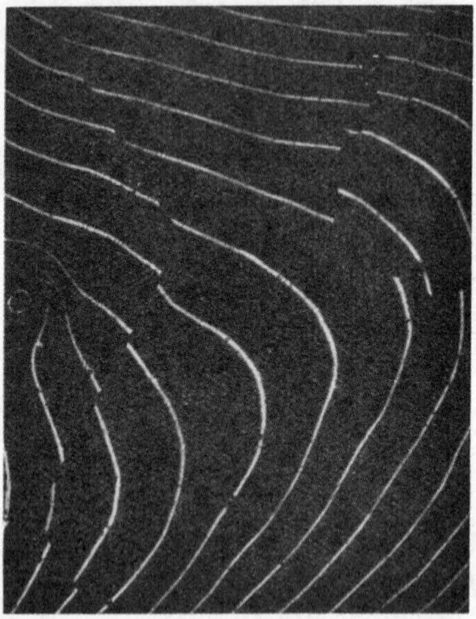

FIG. 101.

Fig. 103 shows a replica illuminated with the mercury yellow doublet instead of the green line. As the fringe pattern is determined by the relative disposition between the silvered surfaces of the specimen and the reference flat, it is natural that the general contours from the mica and from the replica should differ. The important points for comparison are:

(a) the correspondence in the outlines of the cleavage steps;

(b) the heights of the corresponding steps, which are not influenced by flexure of the specimen. (The replica photographs are mirror images of the original mica, and raised features on the latter become depressions on the former.)

The fidelity of reproduction is very clearly shown not only by the numerical data but also by the well-defined small step running in a nearly vertical direction in the right half of Fig. 103. This step, which is only about $\lambda/40$ high, is nevertheless quite readily detectable. Evidence concerning the lower limit of reproducibility is provided by an interesting feature which has emerged from the examination of one particular cleavage line. Upon running along the length of this step on the mica a

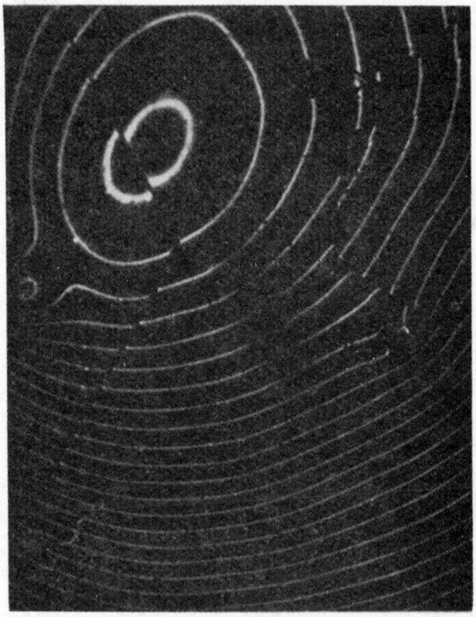

FIG. 102.

discontinuity in step height was found to occur, the height changing from 1,320 to 1,279 A.U. (Each of these values is perhaps in error by 3 A.U.) This change of 41 ± 6 A.U. corresponds to two molecular lattices. The striking fact is that at this point the replica exhibited a corresponding change of 34 ± 6 A.U. Thus notwithstanding the length of the long flexible polymer molecules, the flowing plastic contours the mica so critically that a change in height of only two mica lattices is almost exactly followed.

The characteristic smooth continuity of the mica fringes, formerly shown to indicate that mica cleaves true to a molecular plane over extensive areas, also appears on the replicas. Indeed, despite experience and familiarity extending over several years in the examination of the fringes given by mica, one cannot distinguish between fringes given by micas and those given by replicas, confusing both types completely if descriptive marks are obliterated.

These observations indicate that topographical features as small as 10 A.U. in height are possibly reproducible.

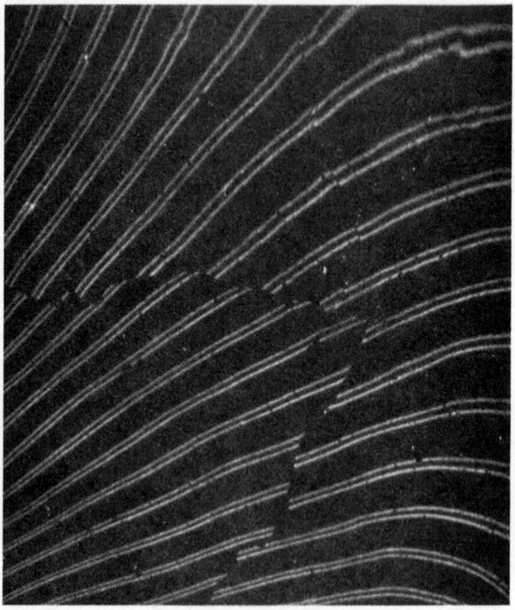

FIG. 103.

Observation shows that a replica grating retains its spacing unchanged over a period of months, and this seems to indicate that creep effects, if any, are not serious.

The moulding technique has been applied to the preparation of reproductions from a cleaved calcite surface and from a polished steel surface, the former soft and friable, the latter hard, yet in both these very different cases excellent mouldings were obtained. In the case of the metal many interesting features in connexion with polish were revealed, without errors due to phase change influencing the conclusions drawn.

CHAPTER XIV

INTERFEROMETRIC COLOUR FILTERS AND MONOCHROMATORS

Principle of colour filter

SUPPOSE two silvered flats are pressed close together, enclosing a thin air film of constant thickness. If placed before a spectroscope slit and illuminated with approximately parallel white light, this arrangement passes a number of specific wave-lengths, which appear as sharp fringes, resembling those shown in Figs. 66 and 76 (without birefringence doubling). The separation between the fringes, in wave-numbers, has already been shown to be $\Delta\nu = 1/2t$. With high reflecting coefficients the amount of light transmitted between the orders is quite small. If t is reduced $\Delta\nu$ increases so that the number of fringes in the visible region diminishes. Examples have already been quoted in which no more than six fringes cover the visible, and it is easy to reduce t until there is only one fringe in the visible region. To take a numerical example: let there be a fringe at λ 5,000. A value of t can be adopted such that the fringes on either side are not in the visible. If $t = 1/14{,}000$ cm. = 7,143 A.U., then the fringes are 7,000 cm.$^{-1}$ apart and appear at $\lambda\lambda$ 3,700, 7,700, neither of which is in the visible. Thus the optical separation of the surfaces must be of the order of a light-wave (red light) to produce an interference film which will behave as a filter.

In this region of λ 5,000 light is transmitted in the form of a wave-length band extending over the effective fringe width. With a reflectivity $R = 0{\cdot}94$ the fringe half-width is about 1/50th of an order, i.e. 140 cm.$^{-1}$, which in this region corresponds to 35 A.U. It can readily be shown that the 'tenth width' is three times the 'half-width', the tenth width being defined as the width between points on the wings which are one-tenth of the intensity of the central maximum. Thus effectively the whole transmission is encompassed within a band 100 A.U. wide.

Some properties of the filter

The graphs in Fig. 104 (adapted from those computed by Donaldson†) show the way in which the number of transmitted fringes depends upon the optical thickness. The selected wave-length range is 4,000–8,000 A.U. and the traverse of the inter-

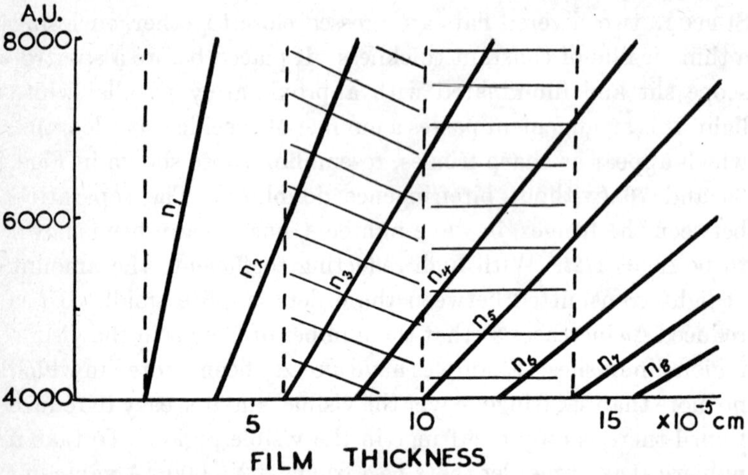

FIG. 104.

ference fringes across the wave-lengths scale with change of film thickness is shown by the straight lines n_1, n_2, n_3,....

Consider the first-order line n_1. Starting with t zero, no fringe becomes visible until $t = 2\times 10^{-5}$ cm. (0·4 of a light-wave at 5,000 A.U.). At this point a blue fringe appears at 4,000 A.U., and as t increases to 4×10^{-5} cm. this single fringe moves across the field up to 8,000 A.U. At this point a second order n_2 appears at 4,000 A.U. and traverses the field as a single fringe until it reaches 6,000 A.U., when the third order at λ 4,000 (n_3) appears, the thickness being 6×10^{-5} cm. From here onwards until $t = 10\times 10^{-5}$ cm. two fringes are visible, either n_2 plus n_3 or n_3 plus n_4. It is clear that from $t = 10\times 10^{-5}$ to 14×10^{-5} three orders are visible between n_3 and n_6 and so on. The t range for a pure colour-filter lies only between 2 and 6×10^{-5} cm. These separations are optical paths and include the change of

† W. K. Donaldson, Ph.D. thesis, Manchester University, 1946.

phase at reflection at the metal surfaces. The regions enclosing successively increasing numbers of fringes are indicated thus:

1. One fringe . . . No shading.
2. Two fringes . . . Cross hatching.
3. Three fringes . . . Horizontal hatching.

Monochromators

The manner in which the colour filters are produced will be described later, and it will be assumed for the purpose of discussion that the desirable thickness t can be obtained. It is clear from Fig. 104 that a transmission band at, say, 5,000 A.U. can be produced either with a film thickness of $2{\cdot}5 \times 10^{-5}$ cm., giving first order, or a thickness 5×10^{-5} cm. giving second order. The second fringe will be the sharper, for in the first case the next fringe appears at 2,500 A.U. and in the second case at 3,333 A.U. Since $\Delta\nu$ separation between fringes is much less, the fringe width in Ångströms is also proportionately less.

Thus, it is seen that the fringe width depends upon the order selected—the higher the order the narrower the fringe. This is self-evident, for if there are, for example, 21 fringes between 4,000 and 8,000 A.U., then $\Delta\nu = 625$ cm.$^{-1}$, making the fringe half-width only $12\frac{1}{2}$ A.U. It is possible by this means to monochromatize a spectrum which has only a few lines in it, such as the mercury spectrum. It is relatively difficult to separate yellow from green lines by ordinary filter methods, and extremely difficult to separate mercury yellow lines from each other by this means, since they are only 20 A.U. apart. For such a purpose costly monochromators must be used. On the other hand, if a multiple-beam filter can be selected to pass one fringe, with fringe width of the order of 10 A.U. or less, then it will cut out the other almost completely. Indeed, by suitable selection of t, it is possible to cut out the green as well as one of the yellow lines, since only those wave-lengths are transmitted which coincide with the intercepts of the lines $n_1, n_2, n_3, \ldots$, made by the vertical line corresponding to the value of t selected.

Thus by adopting the correct t value, a simple inexpensive monochromator can be made for a specific purpose, provided the source to be examined has not many lines. The higher the resolution required, the fewer the lines permitted. For example, it is relatively quite easy to separate the sodium D lines given by a flame or sodium lamp, since there are no other lines of appreciable intensity. A suitably selected piece of mica can be employed for the purpose, for with this, fringes of but a fraction of one Ångström wide can be obtained.

It would appear at first that the problem of securing the *exact* t value is a matter of difficulty. In practice this is not so, for the interference is governed by the relation $n\lambda = 2t\cos\phi$, so that $t\cos\phi$ determines the position of the transmission. By simply tilting the filter the position of transmission can be suitably moved, but only in one direction, namely, towards the blue, since any deviation from the normal reduces $t\cos\phi$.

Compound filters

If a considerable loss in light is not a serious drawback it is possible to make use of a *compound filter* in a manner recalling the use of the compound Fabry-Perot interferometer in high-resolution spectroscopy. The principle is illustrated by Fig. 105. Let A be the fringes passed by a thick filter; as there are many orders each fringe is consequently sharp. Let B be the fringes passed by a thinner filter, the orders being few in number and relatively broad. These conditions can easily be achieved experimentally by using, for example, two pieces of mica of different thicknesses. If the two filters A and B are placed behind each other a composite filter is formed and the transmission is that shown at C (the intensity is of course reduced by the double-absorption effect).

The result of the combination is that the transmitted fringes have at least the sharpness of A with the big dispersion of B. The sharpness is in fact better than that of A, and calculation shows that the fringe width of C is improved upon that of A by a factor $\{K/(1+K)\}^{\frac{1}{2}}$, in which K is ratio of the thicknesses A/B. In the case illustrated this factor is $(\frac{3}{4})^{\frac{1}{2}}$, i.e. a 15 per cent. improvement.

The diagram presupposes an exact integral ratio of A and B.

This may not be the case so that in practice one finds that coincidence can be made exact for one fringe X, but is only imperfect for others. This leads to (a) reduction in intensity in the regions of imperfect coincidence, and (b) the possibility of weak 'ghost' transmission in these same regions. Neither defect

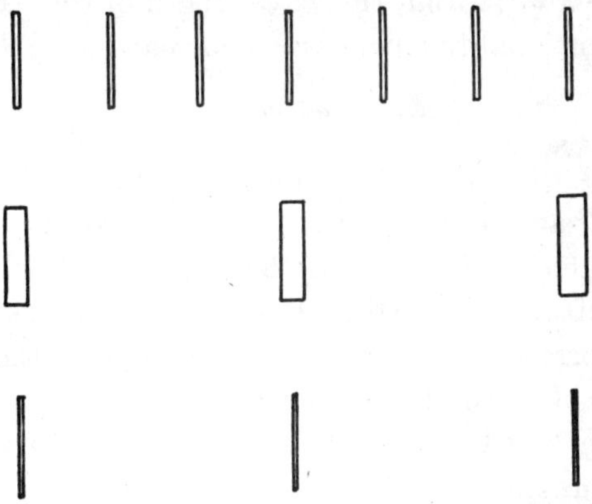

FIG. 105.

need be of any consequence if the region X is that for which the filters are required. A slight relative tilt between A and B permits adjustment for exact coincidence at the point X.

In the next section, dealing with the preparation of the filters, a procedure will be indicated whereby a compound filter could be used with a considerable increase in transmission intensity.

Production of colour filters

It occurred to W. K. Donaldson and J. E. Myers that the principles developed above by the writer could be put into practice by using an evaporated film of a glass-like cryolite—as the thin separator between the two silvered surfaces—instead of using the air film employed originally. J. Brossel improved on this by the addition of a little aluminium oxide to the cryolite. The filter is prepared without breaking the vacuum, as follows. The vacuum evaporating plant has two filaments, one holding silver, the other the cryolite mixture. A piece of glass is used as the base, *an optical flat not being essential.* A silver film is evaporated on to the glass and this contours all the irregularities.

A controlled thickness of cryolite is then evaporated on to the silver, the thickness being controlled by observing the deposition formed upon an unsilvered glass surface, the change in colour and intensity of the reflected image of a pea lamp being noted. (This is a two-beam interference effect but is of sufficient sensitivity.) Since $\lambda = (2/n)\mu t$, the wave-length of the band-pass is directly proportional to thickness, such that

$$d\lambda = (2/n)\mu\, dt.$$

Thus for first-order fringes a change of 100 A.U. in the wave-length of the band ($d\lambda$) is produced by a change of 33 A.U. in the thickness (dt) if $\mu = 1{\cdot}5$. If the transmitted colour is not to have a half-width exceeding that postulated, the thickness must be uniform to less than 30 A.U. Adopting the criterion of a tolerance of 1/5th of a fringe width, this means uniformity in t to within 6 A.U. This occurs locally over a reasonable extension in practice.

After deposition of the cryolite, a second silver film is deposited, and the filter is complete.

When a white-light source (or skylight) is viewed, the purity of the colour transmitted is striking, and filters show uniformity in tint over quite a considerable area (some square centimetres). It is to be noted that as the filter is not formed on glass of good optical quality, the observed colour uniformity is an index of the perfection of contour of the surface by the films. This fact proves two important points. (1) Despite the distorted nature of the surface of the ordinary glass base, the film thickness is uniform. Thus the first silver layer must have perfectly contoured the glass, then the cryolite must have contoured the silver, and finally the second silver layer must have contoured the cryolite. As a result of this the filter thickness is constant. Since any variations in angle of incidence due to the surface topography of the glass are second-order effects, the final interference film is optically as good as if deposited on an optical flat. (2) Not only is the thickness constant, but these observations prove, too, that the deposition is uniform over the surface to within 6 A.U. Further, any serious discrepancy in thickness would produce a change in colour tint easily observable. These observations afford important additional evidence for the view, repeatedly stressed formerly, that the evaporated silver film does in practice truly contour the topography of the surface on to which it condenses.

Optical properties of the filters

A microphotometer tracing of a typical filter showing a single transmission in the visible (second order) is shown in Fig. 106.

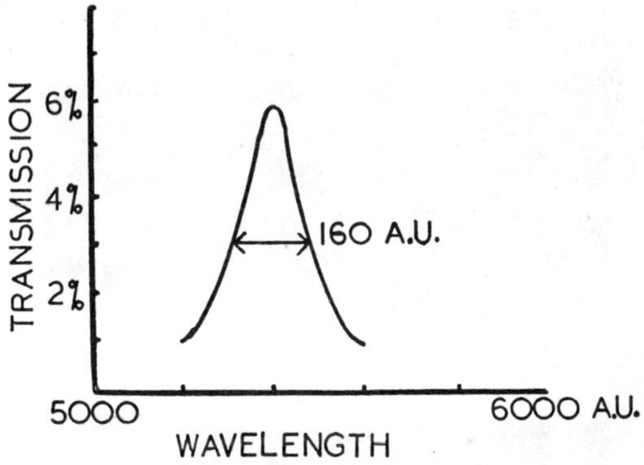

FIG. 106.

The transmission is low, about 6 per cent. The fringe is asymmetrical because of the variation in reflection coefficient with wave-length, the maximum being at λ 5,400. The half-width is 160 A.U., for the reflection coefficient adopted was 0·90. The calculated value is of the same order.

It has been pointed out that an alteration in the angle of incidence will change the position of the maximum since $t \cos \phi$ diminishes. At the same time, if ϕ increases sufficiently then a new effect appears. The differential polarization phase change at reflection at the silver (see chapter on Newton's rings) leads to a doubling of the fringe. As ϕ increases, the main fringe first moves to the violet, then splits into two, and the component on the blue side moves off more rapidly into the violet, but, as previously explained, becomes weaker. The two fringes are plane polarized, mutually perpendicularly, and the secondary can be completely cut out with a polaroid disk. At 50° incidence

500 A.U. separates the components in the case of the filter of Fig. 106. In varying ϕ to alter the position of the maximum, as discussed previously, it may be noted that up to 5° or even 10° the differential effect is negligible and the fringe width is not seriously affected by it.

Filter spectra

The transmission spectrum of a green filter (λ 5,500) is shown in Fig. 107. This has a narrow band-pass near λ 5,500 and another below λ 4,200.

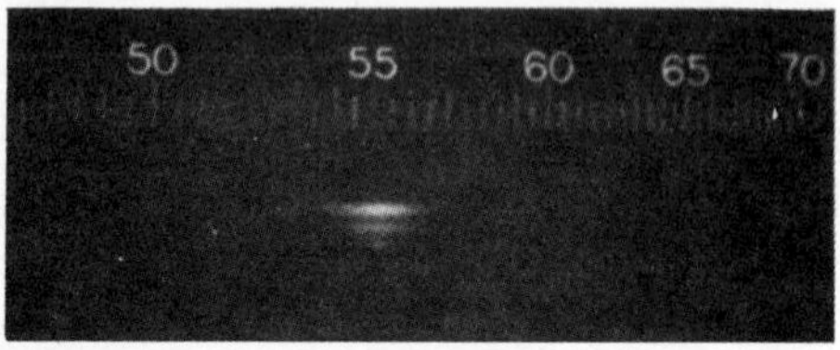

FIG. 107.

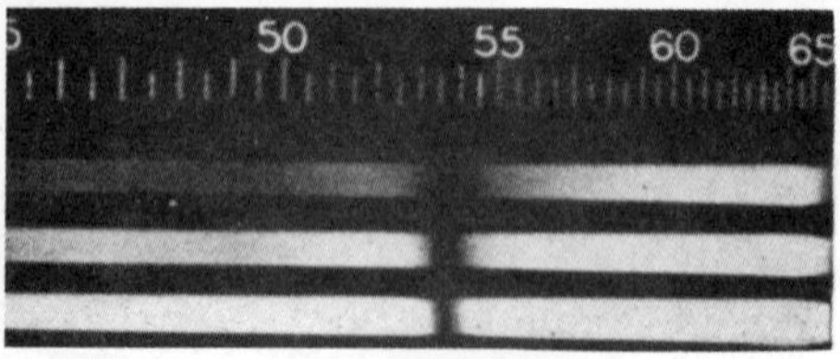

FIG. 108.

From the discussion given on pages 135–50 dealing with reflected fringes, it is clear that the filters should show a complementary reflection spectrum, having narrow dark bands in a bright continuum. Further, such a reflection filter has a high effective luminosity since the light absorption in the film has little effect upon the exposure time. Fig. 108 shows the reflection spectrum corresponding to the transmission spectrum of Fig. 107. It may be noted that the two wave-lengths do not quite coincide, the reason being that in taking the photograph of Fig. 108 the reflected light was not used at exact normal incidence, for convenience in the optical set-up. The reflected fringe appears to be much narrower than the transmitted band. This is partly a photographic illusory effect. If a broad band-pass can be tolerated, a transmission efficiency of perhaps 30 to 50 per cent. can be achieved.

Compound filters

The manner whereby a compound filter is formed from two pieces of doubly silvered mica has already been discussed. The factors affecting the intensity of such an arrangement require further consideration. It has been shown that the fringe intensity given by a multiple-beam interferometer is less than that of the incident light, being reduced by a factor $\{T/(T+A)\}^2$, where T is the transmission and A the absorption. If two identically silvered pieces of mica are set up parallel to each other one might, at first, expect that the transmitted intensity would be again reduced by the same factor so that the combined filter would have a transmission factor $\{T/(T+A)\}^4$. Since $\{T/(T+A)\}^2$ may easily be smaller than 1/10th, it would appear that the combined filter transmission would be less than 1 per cent. of the incident light, possibly much less. This would make its employment prohibitive.

In making the above calculation no consideration has been given to the light reflected between the two filters, and when this is taken into account the transmitted intensity is profoundly modified. A beam emerging from the first filter with intensity P strikes the second filter, and at the first face a fraction RP is reflected. This quantity is not lost since a fraction R^2P is returned from the first filter. This process repeats and the transmission is increased from P to the amount

$$P(1+R^2+R^4+R^6+...).$$

Hence the final transmission increases by a factor $1/(1-R^2)$, which for $R = 0{\cdot}90$ is about 5, and for the higher value $R = 0{\cdot}94$ is about $8\frac{1}{2}$. Surprisingly enough, the combination of two filters does not absorb much more than one, provided they are set parallel.

A still higher efficiency can be achieved by the following cryolite compound filter. A composite filter consisting of silver-cryolite-silver-cryolite-silver has two cryolite films but only three silver films. The absorption is thus reduced, and the emerging peak intensity may in fact be higher than that of a simple filter with two silver films. This arrangement has not yet been tried but offers promise.

Further applications

The colour-filter technique described above should be applicable to the study of monolayers, both organic and inorganic. Such thin films as are formed under these conditions cannot easily be examined by either Fizeau fringes or fringes of equal chromatic order. A colorimetric study might prove profitable in such instances.

CHAPTER XV

NON-LOCALIZED MULTIPLE-BEAM FRINGES

IN this chapter a brief account is given of a method which has been developed for producing non-localized multiple-beam fringes. These were first obtained with a Fabry-Perot interferometer (separation between the plates being 1·5 mm.) and were later applied to the examination of mica cleavage steps. Up to the present this has been their only use with reference to crystal studies. The striking characteristic of these fringes is that they are formed with relatively enormous dispersion by the simplest of means; indeed, circular fringes many *metres* in diameter can readily be obtained without the use of any lens.

A Fabry-Perot interferometer is generally used with an extended source, an essential arrangement if the instrument is placed in the parallel beam. The visible ring system appears only on the area covered by the source image when the latter arrangement is employed, hence the image must extend if several fringes are to be observed. The Fabry-Perot fringes are at infinity and are normally projected on to a screen or photographic plate with a lens, the size of the pattern observed being proportional to the focal length of the image-forming lens.

Despite the long history of the use of the Fabry-Perot interferometer, it had not apparently been noticed that a small source (an approximate point source) can lead to the formation in space of non-localized fringes which extend in cones from the interferometer outwards indefinitely. These cones diverge from the small source and form circular fringes on a screen. A simple mathematical treatment will be given later to show how these conical fringes arise, but even without detailed analysis their origin is clear from the following considerations. A property of the Fabry-Perot interferometer is that in effect it behaves as an 'angular filter'. The positions of the interference maxima are governed by the expression $n\lambda = 2\mu t \cos\phi$. Because of the multiple beams the fringe maxima are sharp and, in effect, appreciable light is only transmitted by the instrument in angular regions close to those values of ϕ for which n is integral. To a crude approximation it can be considered that only the light emerging along the regions close to these integral values of ϕ is strong, light for other angles of emergence being effectively suppressed. The light, therefore, emerges along cones of a definite thickness depending upon the reflecting coefficient of the interferometer mirrors, provided certain conditions relating to the size of the source and the value of t are fulfilled.

If, now, a screen be placed behind the interferometer, circular rings appear upon it. If a certain minimum distance between source and screen is exceeded, the rings formed very closely resemble Fabry-Perot rings. No lenses are needed for the production of these rings since the conical fringes are non-localized, spreading out indefinitely. The farther the screen is removed from the source, the larger are the rings. These rings, formed by the intersection of the non-localized conical fringes with a plane, are not to be confused with the Fabry-Perot rings, which are at infinity and can only be projected on to a screen with the aid of a lens (or seen by accommodating on infinity). The appearance of the two types of rings is, however, very similar, for reasons which will appear in the discussion following.

Size of the rings

With the green mercury line, and t even as large as 1·5 mm., it is easy to produce rings of exceptional diameter on a screen some metres from the source. The first ring can have a diameter of the order of *half a metre* or more, the outer ring diameters being several metres.

The angular diameter of the pth ring can be shown to be $\phi_p = \sqrt{(4p/t)}$, if, for convenience, the case is taken at which the order of interference at the centre is integral. The linear diameters of the rings on a screen are those which would be formed by the same interferometer used in the normal manner of Fabry and Perot when employing a projecting lens with focal length equal to the source-screen distance in the conical fringe observation.

With a thin piece of mica ($t = 1/100$ mm.) the angular

diameter of the first ring (λ 5,461) is almost 0·47 radian. Thus, on a screen placed only *1 metre* from the source, successive rings are produced of diameters about 0·54, 0·75, 0·92, 1·08,... *metres.* But with the intense point-sources now available it is possible to move the screen to a distance of 10 metres, making the first ring diameter 5·4 metres and the separation between, for example, the second and third fringes as great as 170 cm.

Such an abnormally great dispersion is usually not required, but it is of interest to point out that it is available. In practice it is only necessary to place the screen 1/100th of this distance from the source (i.e. 10 cm.) to obtain a suitably large dispersion for mica, and even such dispersion usually far exceeds that available by other methods involving auxiliary instruments.

The theory of the ring formation

The theory of formation of the fringes is as follows. Let it be assumed that a monochromatic point-source S is situated at some distance from the interferometer PP, the plate separation of which is t (see Fig. 109). As a result of the multiple reflections, the effect upon a screen is equivalent to that produced by a series of equidistant point-sources S, S_1, S_2, S_3,..., each $2t$ apart. If the distance L from the source S to the screen is large compared with $2t$, then to a close approximation the intensities of successive beams reaching any point Q fall off geometrically. (If the small reduction in intensity due to increasing distance be taken into account, the net final effect can be considered as equivalent to a slight reduction in the reflection coefficient of the mirrors. This is a refinement which need not be considered here.) The position of the interferometer relative to the source S is immaterial, provided the aperture of the plates suffices to permit the passage of sufficient beams.

Light from the succession of virtual point-sources leads to the formation of circular rings upon the screen. (A corresponding infinite set of virtual images is formed on the other side of the interferometer. These will lead to the formation of a complementary system of conical non-localized fringes expanding outwards to the left of the diagram. These back-reflected fringes are not considered here.) If the lengths of the paths from the successive points S, S_1, S_2, S_3,..., were to increase in arithmetical progression, then the rings formed on the screen would have the intensity distribution of Fabry-Perot fringes. Under suitable conditions the deviations of the retardations from this exact condition are not serious.

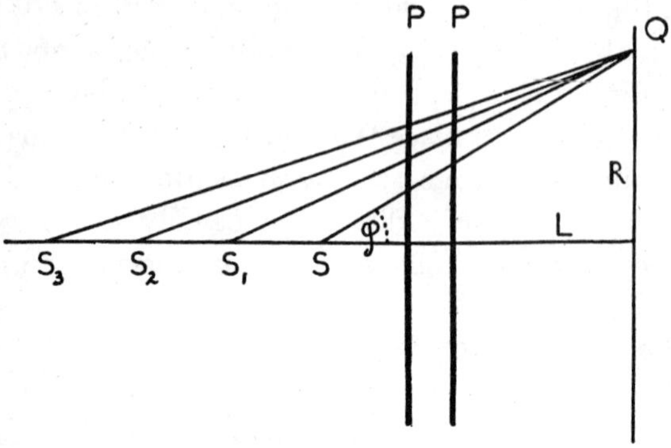

FIG. 109.

The path difference at the point Q between the two rays from S and S_1 can be written as

$$D = \sqrt{(R^2+L^2+4Lt+4t^2)}-\sqrt{(L^2+R^2)}$$

$$= (L^2+R^2)^{\frac{1}{2}}\left\{\left(1+\frac{4Lt+4t^2}{L^2+R^2}\right)^{\frac{1}{2}}-1\right\}$$

$$= (L^2+R^2)^{\frac{1}{2}}\{(1+\alpha)^{\frac{1}{2}}-1\},$$

where
$$\frac{4Lt+4t^2}{L^2+R^2} = \alpha.$$

Expanding the expression containing α to include second-order terms gives

$$D = (L^2+R^2)^{\frac{1}{2}}\left(\frac{\alpha}{2}-\frac{\alpha^2}{8}\right).$$

If SQ makes an angle ϕ with the axis, then

$$D = \frac{L}{\cos\phi}\left\{\left(\frac{2Lt+2t^2}{L^2}\right)\cos^2\phi-2\left(\frac{L^2t^2+2Lt^3+t^4}{L^4}\right)\cos^4\phi\right\}.$$

Powers of t/L greater than the second can be neglected, giving

$$D = \frac{L}{\cos\phi}\left\{\frac{2t}{L}\cos^2\phi + \frac{2t^2}{L^2}(\cos^2\phi - \cos^4\phi)\right\}$$

$$= 2t\cos\phi + 2(t^2/L)\cos\phi\sin^2\phi.$$

The path difference D_n between the first and nth beam is obtained if t is replaced by nt, giving

$$D_n = 2nt\cos\phi + 2n^2(t^2/L)\cos\phi\sin^2\phi.$$

For the first ring $D = 2T - \lambda$ making, closely enough, $\sin^2\phi = \lambda/t$. Thus

$$D_n = 2nt\cos\phi + 2n^2t(\lambda/t)\cos\phi.$$

The amount by which the nth ray lags behind the arithmetical progression value is

$$(n^2-1)\frac{2t\lambda}{L}\cos\phi.$$

As $\cos\phi = \left(1 - \frac{\lambda}{2t}\right)$ very closely, the lag Δ is

$$(n^2-1)\left(\frac{2t\lambda}{L}\right)\left(1 - \frac{\lambda}{2t}\right),$$

i.e.

$$\Delta = (n^2-1)\left(\frac{2t\lambda}{L} - \frac{\lambda^2}{L}\right).$$

The term in λ^2 can be neglected, hence

$$\Delta = (n^2-1)\frac{2t\lambda}{L}.$$

Adopting $t = 1/100$ mm., then even the small source–screen distance of 10 cm. (making $t/L = 10^{-4}$) leads to a phase retardation of only 6×10^{-4} of a wave for the first pair of beams.

A retardation of a complete $\lambda/2$ is only reached for $n = 50$. There is thus almost complete Airy summation, even at 10 cm. distance.

The simple calculations just made refer to the reinforcement of the successive beams with the first beam only. It is necessary for completion of the argument to consider the points of reinforcement for beams such as S_{n+1}, S_{n+2}, S_{n+3},... with S_n, etc. In general, rays from S_n and S_{n+1} reinforce at a point distant $R = 2n\sqrt{(\lambda t)}$ from that at which S and S_1 unite. This is of the same order as the displacement formerly considered. From the known reflection coefficient it is possible to plot the total distribution graphically, and this shows that the fringes are unsymmetrical. However, the distribution of intensity so closely approximates that of the Fabry-Perot distribution, that differences can only be detected in rings extremely far from the centre, of the 200th order or greater.

Further properties

The diameters of the rings are not quite the same as those of the corresponding Fabry-Perot rings, although so close to the latter that differences cannot easily be detected. From expressions already given, it follows that in effect t in the Fabry-Perot formula has changed to $t\{1+(n\lambda/L)\}$ for the non-localized first ring. The quantity λ/L is some $5{\cdot}4\times10^{-6}$. A graphical construction of the distribution of the points of reinforcement of the more important beams gives a centre of gravity not far from the two beams S and S_1. The construction shows that the optical centre of gravity lies in the approximate neighbourhood of $n = 5$. It follows, then, that for the first ring t is effectively increased by only twenty-seven parts in a million, and proportionately less for larger values of L.

The quantity $(2nt/L)$ gives the increase in order of interference for the first ring compared with that of the Fabry-Perot. With the values adopted here this has the numerical values 1/1,000 at the optical centre of gravity. The diameter of the first ring is therefore decreased by 1/1,000th of an order relative to the Fabry-Perot.

For the pth ring t must be replaced by $t\{1+(pn\lambda/L)\}$. Hence the pth ring does not occupy the position of the corresponding ring in the Fabry-Perot system but is displaced by p/1,000th of an order inwards. The rings therefore gradually, almost inappreciably, shrink inwards, and if superposed on the Fabry-Perot system one ring would be overtaken by the 1/1,000th order.

This effect is of little consequence and may be disregarded.

Not only do the diameters of higher order rings change, but in addition a much more obvious effect on the fringe width takes place as p increases. The additional phase retardation is proportional to p, being equal to $(n^2-1)p(2t/L)$ for the nth beam reaching the pth ring. For large values of $p(2t/L)$ the Airy sum condition is only approximately fulfilled and the fringes broaden. Since with the present values $2t/L$ is equal to 1/5,000, the effect due to this is of no consequence even at the 100th ring.

Of importance, however, is the finite extension of the source. It is clear that if the light source is a single disk, each point on the disk produces its own independent ring system, and it follows that the width of every ring is increased because of the finite source dimension. To the natural ring width, which depends upon the reflecting coefficient and the order number of the ring, must be added an amount equal to the diameter of the source.

The fringe width on the screen due to the instrumental intensity distribution is linearly proportional to the source–screen distance, hence the larger the value of L the less important is the influence of the finite source dimension.

Application to examination of mica

For examination with non-localized fringes a mica sheet is silvered on both sides. Instead of using the mica as a simple angular filter placed before a source, as considered in the analytical treatment given above, it is advisable to project an image of the source on to the mica, and by this means the properties of small local areas can be examined. If this is not done, the effects at different areas introduce confusion. A green filter is placed before a mercury arc and an image projected on to an aperture of diameter 1 mm. With a 1 in. microscope objective a reduced, bright image of this circular aperture is projected on to the surface of the mica, and this acts as the point source for the production of the non-localized fringes.

With a thin slip of mica the dispersion is so great that it is often not possible to accommodate more than a single ring on a quarter-plate placed some 10–20 cm. away from the mica. It is then necessary to incline the mica through an appreciable angle to tilt the ring system off-centre, and by this means two or three orders can be accommodated on the plate, an essential condition for measurement.

The first example of applying this technique is shown in Fig. 110, the piece of mica selected showing large uniform tint areas with Fizeau fringes. In the case of mica it is necessary first to plane-polarize the light if birefringence doubling is to be

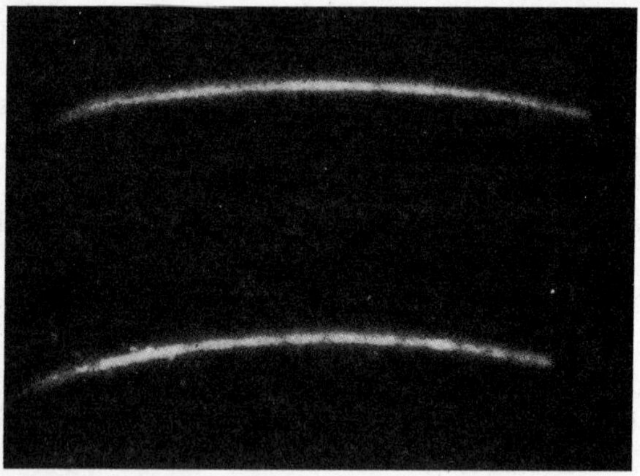

FIG. 110.

avoided. This has been done in this instance. The rings are sharp and the high dispersion is evident, for the reproduction is a contact print with unit magnification.

Despite the tilting to bring the rings off-centre to permit of segments of two rings falling on the plate, the linear distance between orders is 3 cm. It could have been made 30 cm. by moving the photographic plate 2 metres from the mica instead of the 20 cm. used. The light intensity easily permitted this. It will be noticed that the ring segments chosen are only slightly curved, indicating how large are the complete rings.

With green mercury light a change in mica thickness of 1,700 A.U. is needed to displace the rings through a single order. Since it is evident from the fringe sharpness that changes of 1/100th order can be measured, one can expect to detect a step of at least 17 A.U., probably 12 A.U. The sensitivity is therefore the same as that of sharp Fizeau fringes and fringes of equal chromatic order. It is surpassed only by the uniform tint technique employed with crossed fringes.

Cleavages

A piece of mica was selected which had on it a few long cleavage lines. The image of the source was adjusted to fall across one of these lines and the non-localized fringes thus

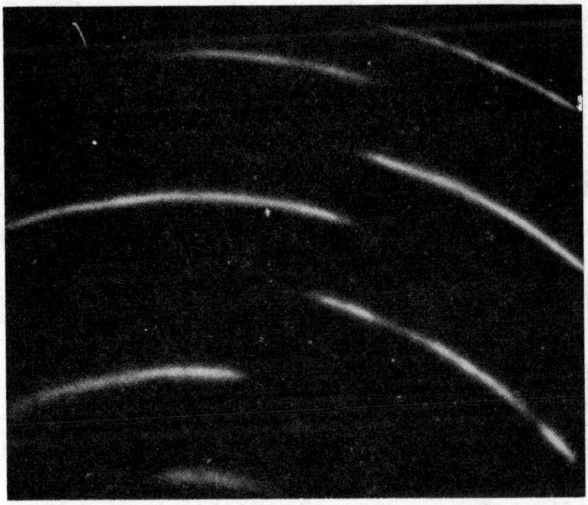

FIG. 111.

obtained are shown in Fig. 111. A discontinuity appears in the fringe pattern because of the cleavage step. In this region there exist effectively two Fabry-Perot interferometers side by side, differing in gap only by the height of the cleavage step. The incident converging cone of light forming the 'point' source is divided by the cleavage line so that on one side of the line some half of the cone is available and therefore only half the field of view is covered by this system. On the other side of the cleavage line appears the remaining part of the ring system, from the adjacent Fabry-Perot interferometer, with a displacement in order corresponding to the difference in thickness.

From the displacement of the fringe, which can be evaluated by the McNair approximation described on page 42, the value of the step can be calculated. One photograph, however, is insufficient to determine the value of the step since there is no indication of the direction, i.e. whether the right-hand fringes have moved up a fraction x or moved down a fraction $1-x$, or whether the step is greater than one order. Two methods can be employed for resolving the ambiguity. The first method is laborious. Different wave-lengths are used and then Benoit's method of exact fractions can be applied. It is somewhat out of place here to go into the details of this procedure since it is well known and is the basis of the method of determining a Fabry-Perot gap with precision when making accurate wave-length determinations. This method requires further a knowledge of the dispersion and the refractive indices for the wave-lengths employed.

An obvious simple method for determining the correct allocation of orders is to compare the non-localized fringes with corresponding fringes of equal chromatic order. It may be considered that little is gained by using non-localized fringes, but this is not so, for the necessary fringes of equal chromatic order can be produced with a low-dispersion visual spectroscope, the non-localized fringes being used for the high-dispersion and precision measurements.

High-dispersion fringes of equal chromatic order may require a costly dispersion spectrograph and this may not be available. In this case the non-localized fringes can be used. Furthermore, the dispersion of the fringes of equal chromatic order is primarily determined by the thickness of the interference film. A film 1/10 mm. thick cannot produce high-dispersion fringes of equal chromatic order, even with a large prism spectrograph (say, Hilger E.I. with glass train), since the orders are only about 33 cm.$^{-1}$ apart. But with such a film it is still easy to obtain great dispersion with the non-localized fringes.

Thus the high dispersion of the non-localized fringes is a valuable property, particularly for thick samples. The technique is also to be recommended if a big spectrograph is not available.

The birefringence

When the polarizing device is removed from the source the appearance changes from that in Fig. 111 to that in Fig. 112.

The birefringence leads to a doubling of the fringes, one set circular, the other slightly elliptical. A closely similar effect was first indicated in work of Rayleigh (1906) on the Fabry-Perot rings (at infinity) given by doubly silvered mica. Rayleigh's fringes were crude, yet he found intensity variations in the ring

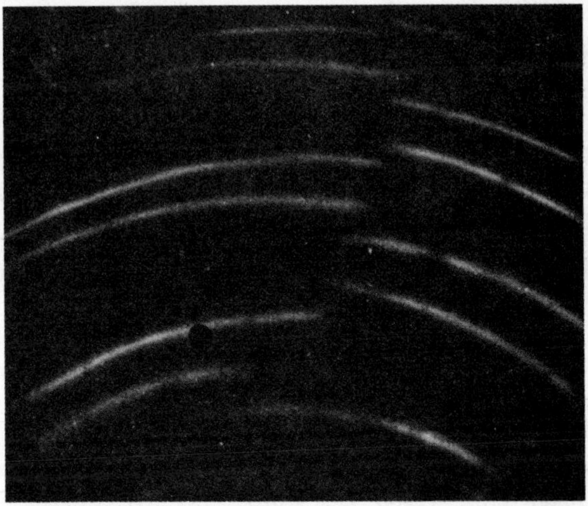

FIG. 112.

system, due, as he showed, to superposed unresolved circular and elliptical ring patterns arising from birefringence. More recently Pfund (1942) has clearly established the existence of the circular and elliptical Fabry-Perot rings and has discussed their origin.

The birefringence effect shown here for the non-localized fringes is to be identified with that reported by Pfund for the Fabry-Perot rings, but of course very high dispersions are available if need be. It is true that Fabry-Perot rings with thin mica sheets also have high dispersion, yet an added advantage of the non-localized fringes is that a continuously variable dispersion is to hand with the further added advantage that there are no lens distortions.

Further examples

Fig. 113 shows further examples of non-localized fringes (without birefringence complications) with a different mica sample, and this illustrates possible applications. The fringes were given by a sample of mica which showed a multiplicity of small cleavage areas. In Fig. 113*a* the small image of the source covers an area which contains a large number of discrete strips extending lengthwise in a horizontal direction. Several of these strips (six) fall together within the small dimension of the source

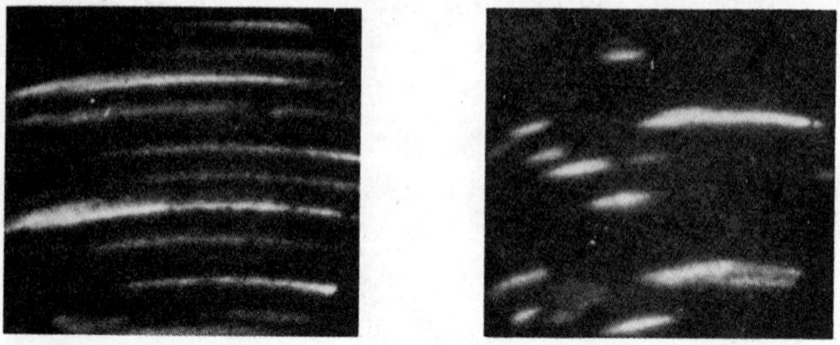

(*a*) (*b*)

FIG. 113.

image (say, 1/5 mm.). From the horizontal extension of the fringes it is clear that the strips extend over sufficient length to encompass a considerable angular range for each strip. (This length need not exceed the source-image diameter for this purpose.)

Fig. 113*b* was obtained by moving the source image to another area. In this case the narrow strips possibly run vertically, but in any event their extension horizontally covers only a small part of the source diameter. When this sample of mica was moved across the field the source image traversed different areas, and a bewildering flicker of fringes passed over the screen, indicating the structural complexity.

Further applications

The non-localized fringes have as yet hardly been applied, but it is clear that the high dispersion available offers interesting possibilities. The difficulty as to order allocation and step direction does not always exist. Thus suppose the technique is applied to the examination of monolayers put down on a silver surface, or to the measurement of the thickness of a very thin layer of metal. In both cases it is often known that the film thickness is a small fraction of a light-wave, and the direction of the step can also be self-evident. Thus we have at our disposal a simple high-dispersion method for measuring such steps, without auxiliary instruments.

A further possible use is for the evaluation of very small birefringence effects. At the centre of the ring system $n\lambda = 2\mu t$, hence $\lambda\,dn = 2t\,d\mu$, giving $d\mu = dn(\lambda/2t)$, in which dn is the fractional order change at the centre. But if the tilt on the whole system is not large, this is the fractional order between any fringe pair, hence it is easy to obtain $d\mu$. The thickness t can be measured by any of several methods, including, if required, optical methods already discussed.

The value of dn observed is proportional to t, thus the thickest specimen possible should be used if $d\mu$ is very small. Suppose a thin crystal, say, 1/10 mm. thick, is available, then for, say, λ 5,000, $dn = d\mu \times 400$. Thus a birefringence as small as

$$d\mu = 0{\cdot}0001$$

will produce a fringe doubling with $dn = 1/25$th of an order and this can be resolved. So that even for so thin a crystal a low birefringence of only 1×10^{-4} can be measured, and the thicker the crystal, the greater the sensitivity. Fine fringes are still given by a 1 mm. gap, enabling in this case $d\mu = 1 \times 10^{-5}$ to be measured, but the source–screen distance should then be of the order of half a metre or so. Fig. 112 shows clearly how easy it is to observe birefringence even with a mica sheet 1/100 mm. thick.

A further useful application is the testing of the perfection of parallelism, i.e. the testing of uniformity of film thickness. Since the fringes are non-localized they can be viewed anywhere in space with a lens or low-power microscope. (When the eye alone moves across the field no fringes are seen, but the light intensity flickers as the eye crosses over the maxima and minima.) If the fringes are observed and the interference film moved across the field of view so that the source image scans the film, then no movement of fringes can be seen if the optical thickness in the film is constant. If the thickness changes, the fringes move, expanding out if the thickness increases. Since $n\lambda = 2\mu t$, then

$$dt = \frac{dn\,\lambda}{2\mu} = dn\frac{\lambda}{3} \quad \text{(closely enough).}$$

A change of fringe position by a whole order is produced for $dt = \lambda/3$. It is probable that a change of 1/100th of an order could be detected (1/3 mm. shift in the cases reproduced) which corresponds to $dt = 17$ A.U.

This method for measuring uniformity in thickness is in fact that proposed long ago by Fabry for the adjustment of the Fabry-Perot interferometer. The aim there is to adjust two plates parallel (i.e. constant gap) and deviations are shown up by fringe displacements observed with a *telescope* set on infinity. In the present case the closely analogous arrangement holds, but with a *microscope* set on the non-localized fringes.

The procedure might be useful in the case of deposited multiple monolayers, and affords an alternative procedure to those described on pages 182–3. With such layers t will in general be so small that the source–screen distance will be no more than a few centimetres.

BIBLIOGRAPHY

1. FIZEAU. *Compt. Rend.* **54,** 1237 (1862).
2. HAIDINGER. *Pogg. Ann.* **77,** 217 (1849).
3. TWYMAN and GREEN. Brit. Patent 103832 (1916).
4. FABRY and PEROT. *Ann. de Chim. et de Phys.* **12,** 459 (1897).
5. BENOIT, FABRY, and PEROT. *Trav. et Mém. Bureau Inter. de Poids et Mesures,* **15** (1913).
6. Personal communication from W. E. Williams.
7. AIRY. *Math. Tracts,* p. 381 (1831).
8. BOULOUCH. *Journ. de Physique,* **5,** 789 (1906).
9. BURGER and VAN CITTERT. *ZS. Physik,* **44,** 58 (1927).
10. TOLANSKY. *Phil. Mag.* **35,** 120 (1944), 179 (1944), et seq.
11. —— *Proc. Roy. Soc.* A, **184,** 41 and 51 (1945).
12. —— Ibid. **186,** 261 (1946).
13. FABRY. *Revue d'Optique,* **1,** 445 (1922).
14. TOLANSKY. *High Resolution Spectroscopy* (Methuen, 1946).
15. RITSCHL. *ZS. Physik,* **69,** 578 (1931).
16. STRONG. *Modern Laboratory Practice* (1938).
17. TOLANSKY. *Phil. Mag.* **35,** 120 (1944).
18. FEUSSNER. *Gehrckes Handb. d. Phys. Optik,* vol. i (1927).
19. McNAIR. *Phil. Mag.* **2,** 613 (1926).
20. McLAURIN. *Proc. Roy. Soc.* A, **78,** 206 (1906).
21. SIEGBAHN. *M. Ak. Matem. Astro. Fysik.* **23,** 12 (1932).
22. KAYSER. *Indust. Diamond Rev.* **4** (Jan. 1944).
23. BUCKLEY. *ZS. f. Kristallographie,* **8,** 410 (1934).
24. TOLANSKY. *Proc. Roy. Soc.* A, **184,** 41 (1945).
25. BROSSEL. *Nature,* **157,** 623 (1946).
26. TOLANSKY. *Proc. Roy. Soc.* A, **184,** 51 (1945).
27. —— and WILCOCK. *Nature,* **157,** 583 (1946).
28. —— *Phil. Mag.* **36,** 225 (1945).
29. —— Ibid. 236 (1945).
30. —— and KHAMSAVI. *Nature,* **157,** 661 (1946).

INDEX

SOME DOVER SCIENCE BOOKS

CHANCE, LUCK AND STATISTICS: THE SCIENCE OF CHANCE,
Horace C. Levinson

Theory of probability and science of statistics in simple, non-technical language. Part I deals with theory of probability, covering odd superstitions in regard to "luck," the meaning of betting odds, the law of mathematical expectation, gambling, and applications in poker, roulette, lotteries, dice, bridge, and other games of chance. Part II discusses the misuse of statistics, the concept of statistical probabilities, normal and skew frequency distributions, and statistics applied to various fields—birth rates, stock speculation, insurance rates, advertising, etc. "Presented in an easy humorous style which I consider the best kind of expository writing," Prof. A. C. Cohen, Industry Quality Control. Enlarged revised edition. Formerly titled *The Science of Chance*. Preface and two new appendices by the author. xiv + 365pp. 5⅜ x 8. 21007-3 Paperbound $2.00

BASIC ELECTRONICS,
prepared by the U.S. Navy Training Publications Center

A thorough and comprehensive manual on the fundamentals of electronics. Written clearly, it is equally useful for self-study or course work for those with a knowledge of the principles of basic electricity. Partial contents: Operating Principles of the Electron Tube; Introduction to Transistors; Power Supplies for Electronic Equipment; Tuned Circuits; Electron-Tube Amplifiers; Audio Power Amplifiers; Oscillators; Transmitters; Transmission Lines; Antennas and Propagation; Introduction to Computers; and related topics. Appendix. Index. Hundreds of illustrations and diagrams. vi + 471pp. 6½ x 9¼.
61076-4 Paperbound $2.95

BASIC THEORY AND APPLICATION OF TRANSISTORS,
prepared by the U.S. Department of the Army

An introductory manual prepared for an army training program. One of the finest available surveys of theory and application of transistor design and operation. Minimal knowledge of physics and theory of electron tubes required. Suitable for textbook use, course supplement, or home study. Chapters: Introduction; fundamental theory of transistors; transistor amplifier fundamentals; parameters, equivalent circuits, and characteristic curves; bias stabilization; transistor analysis and comparison using characteristic curves and charts; audio amplifiers; tuned amplifiers; wide-band amplifiers; oscillators; pulse and switching circuits; modulation, mixing, and demodulation; and additional semiconductor devices. Unabridged, corrected edition. 240 schematic drawings, photographs, wiring diagrams, etc. 2 Appendices. Glossary. Index. 263pp. 6½ x 9¼. 60380-6 Paperbound $1.75

GUIDE TO THE LITERATURE OF MATHEMATICS AND PHYSICS,
N. G. Parke III

Over 5000 entries included under approximately 120 major subject headings of selected most important books, monographs, periodicals, articles in English, plus important works in German, French, Italian, Spanish, Russian (many recently available works). Covers every branch of physics, math, related engineering. Includes author, title, edition, publisher, place, date, number of volumes, number of pages. A 40-page introduction on the basic problems of research and study provides useful information on the organization and use of libraries, the psychology of learning, etc. This reference work will save you hours of time. 2nd revised edition. Indices of authors, subjects, 464pp. 5⅜ x 8.
60447-0 Paperbound $2.75

THE RISE OF THE NEW PHYSICS (formerly THE DECLINE OF MECHANISM), *A. d'Abro*

This authoritative and comprehensive 2-volume exposition is unique in scientific publishing. Written for intelligent readers not familiar with higher mathematics, it is the only thorough explanation in non-technical language of modern mathematical-physical theory. Combining both history and exposition, it ranges from classical Newtonian concepts up through the electronic theories of Dirac and Heisenberg, the statistical mechanics of Fermi, and Einstein's relativity theories. "A must for anyone doing serious study in the physical sciences," *J. of Franklin Inst.* 97 illustrations. 991pp. 2 volumes.

20003-5, 20004-3 Two volume set, paperbound $5.50

THE STRANGE STORY OF THE QUANTUM, AN ACCOUNT FOR THE GENERAL READER OF THE GROWTH OF IDEAS UNDERLYING OUR PRESENT ATOMIC KNOWLEDGE, *B. Hoffmann*

Presents lucidly and expertly, with barest amount of mathematics, the problems and theories which led to modern quantum physics. Dr. Hoffmann begins with the closing years of the 19th century, when certain trifling discrepancies were noticed, and with illuminating analogies and examples takes you through the brilliant concepts of Planck, Einstein, Pauli, de Broglie, Bohr, Schroedinger, Heisenberg, Dirac, Sommerfeld, Feynman, etc. This edition includes a new, long postscript carrying the story through 1958. "Of the books attempting an account of the history and contents of our modern atomic physics which have come to my attention, this is the best," H. Margenau, Yale University, in *American Journal of Physics*. 32 tables and line illustrations. Index. 275pp. 5⅜ x 8.

20518-5 Paperbound $2.00

GREAT IDEAS AND THEORIES OF MODERN COSMOLOGY, *Jagjit Singh*

The theories of Jeans, Eddington, Milne, Kant, Bondi, Gold, Newton, Einstein, Gamow, Hoyle, Dirac, Kuiper, Hubble, Weizsäcker and many others on such cosmological questions as the origin of the universe, space and time, planet formation, "continuous creation," the birth, life, and death of the stars, the origin of the galaxies, etc. By the author of the popular *Great Ideas of Modern Mathematics*. A gifted popularizer of science, he makes the most difficult abstractions crystal-clear even to the most non-mathematical reader. Index. xii + 276pp. 5⅜ x 8½. 20925-3 Paperbound $2.50

GREAT IDEAS OF MODERN MATHEMATICS: THEIR NATURE AND USE, *Jagjit Singh*

Reader with only high school math will understand main mathematical ideas of modern physics, astronomy, genetics, psychology, evolution, etc., better than many who use them as tools, but comprehend little of their basic structure. Author uses his wide knowledge of non-mathematical fields in brilliant exposition of differential equations, matrices, group theory, logic, statistics, problems of mathematical foundations, imaginary numbers, vectors, etc. Original publications, appendices. indexes. 65 illustr. 322pp. 5⅜ x 8. 20587-8 Paperbound $2.25

THE MATHEMATICS OF GREAT AMATEURS, *Julian L. Coolidge*

Great discoveries made by poets, theologians, philosophers, artists and other non-mathematicians: Omar Khayyam, Leonardo da Vinci, Albrecht Dürer, John Napier, Pascal, Diderot, Bolzano, etc. Surprising accounts of what can result from a non-professional preoccupation with the oldest of sciences. 56 figures. viii + 211pp. 5⅜ x 8½. 61009-8 Paperbound $2.00

FIVE VOLUME "THEORY OF FUNCTIONS" SET BY KONRAD KNOPP

This five-volume set, prepared by Konrad Knopp, provides a complete and readily followed account of theory of functions. Proofs are given concisely, yet without sacrifice of completeness or rigor. These volumes are used as texts by such universities as M.I.T., University of Chicago, N. Y. City College, and many others. "Excellent introduction . . . remarkably readable, concise, clear, rigorous," *Journal of the American Statistical Association.*

ELEMENTS OF THE THEORY OF FUNCTIONS,
Konrad Knopp
This book provides the student with background for further volumes in this set, or texts on a similar level. Partial contents: foundations, system of complex numbers and the Gaussian plane of numbers, Riemann sphere of numbers, mapping by linear functions, normal forms, the logarithm, the cyclometric functions and binomial series. "Not only for the young student, but also for the student who knows all about what is in it," *Mathematical Journal.* Bibliography. Index. 140pp. 5⅜ x 8. 60154-4 Paperbound $1.50

THEORY OF FUNCTIONS, PART I,
Konrad Knopp
With volume II, this book provides coverage of basic concepts and theorems. Partial contents: numbers and points, functions of a complex variable, integral of a continuous function, Cauchy's integral theorem, Cauchy's integral formulae, series with variable terms, expansion of analytic functions in power series, analytic continuation and complete definition of analytic functions, entire transcendental functions, Laurent expansion, types of singularities. Bibliography. Index. vii + 146pp. 5⅜ x 8. 60156-0 Paperbound $1.50

THEORY OF FUNCTIONS, PART II,
Konrad Knopp
Application and further development of general theory, special topics. Single valued functions. Entire, Weierstrass, Meromorphic functions. Riemann surfaces. Algebraic functions. Analytical configuration, Riemann surface. Bibliography. Index. x + 150pp. 5⅜ x 8. 60157-9 Paperbound $1.50

PROBLEM BOOK IN THE THEORY OF FUNCTIONS, VOLUME 1.
Konrad Knopp
Problems in elementary theory, for use with Knopp's *Theory of Functions,* or any other text, arranged according to increasing difficulty. Fundamental concepts, sequences of numbers and infinite series, complex variable, integral theorems, development in series, conformal mapping. 182 problems. Answers. viii + 126pp. 5⅜ x 8. 60158-7 Paperbound $1.50

PROBLEM BOOK IN THE THEORY OF FUNCTIONS, VOLUME 2,
Konrad Knopp
Advanced theory of functions, to be used either with Knopp's *Theory of Functions,* or any other comparable text. Singularities, entire & meromorphic functions, periodic, analytic, continuation, multiple-valued functions, Riemann surfaces, conformal mapping. Includes a section of additional elementary problems. "The difficult task of selecting from the immense material of the modern theory of functions the problems just within the reach of the beginner is here masterfully accomplished," *Am. Math. Soc.* Answers. 138pp. 5⅜ x 8.
60159-5 Paperbound $1.50

NUMERICAL SOLUTIONS OF DIFFERENTIAL EQUATIONS,
H. Levy & E. A. Baggott

Comprehensive collection of methods for solving ordinary differential equations of first and higher order. All must pass 2 requirements: easy to grasp and practical, more rapid than school methods. Partial contents: graphical integration of differential equations, graphical methods for detailed solution. Numerical solution. Simultaneous equations and equations of 2nd and higher orders. "Should be in the hands of all in research in applied mathematics, teaching," *Nature*. 21 figures. viii + 238pp. 5⅜ x 8. 60168-4 Paperbound $1.85

ELEMENTARY STATISTICS, WITH APPLICATIONS IN MEDICINE AND THE BIOLOGICAL SCIENCES, *F. E. Croxton*

A sound introduction to statistics for anyone in the physical sciences, assuming no prior acquaintance and requiring only a modest knowledge of math. All basic formulas carefully explained and illustrated; all necessary reference tables included. From basic terms and concepts, the study proceeds to frequency distribution, linear, non-linear, and multiple correlation, skewness, kurtosis, etc. A large section deals with reliability and significance of statistical methods. Containing concrete examples from medicine and biology, this book will prove unusually helpful to workers in those fields who increasingly must evaluate, check, and interpret statistics. Formerly titled "Elementary Statistics with Applications in Medicine." 101 charts. 57 tables. 14 appendices. Index. vi + 376pp. 5⅜ x 8. 60506-X Paperbound $2.25

INTRODUCTION TO SYMBOLIC LOGIC,
S. Langer

No special knowledge of math required — probably the clearest book ever written on symbolic logic, suitable for the layman, general scientist, and philosopher. You start with simple symbols and advance to a knowledge of the Boole-Schroeder and Russell-Whitehead systems. Forms, logical structure, classes, the calculus of propositions, logic of the syllogism, etc. are all covered. "One of the clearest and simplest introductions," *Mathematics Gazette*. Second enlarged, revised edition. 368pp. 5⅜ x 8. 60164-1 Paperbound $2.25

A SHORT ACCOUNT OF THE HISTORY OF MATHEMATICS,
W. W. R. Ball

Most readable non-technical history of mathematics treats lives, discoveries of every important figure from Egyptian, Phoenician, mathematicians to late 19th century. Discusses schools of Ionia, Pythagoras, Athens, Cyzicus, Alexandria, Byzantium, systems of numeration; primitive arithmetic; Middle Ages, Renaissance, including Arabs, Bacon, Regiomontanus, Tartaglia, Cardan, Stevinus, Galileo, Kepler; modern mathematics of Descartes, Pascal, Wallis, Huygens, Newton, Leibnitz, d'Alembert, Euler, Lambert, Laplace, Legendre, Gauss, Hermite, Weierstrass, scores more. Index. 25 figures. 546pp. 5⅜ x 8.
20630-0 Paperbound $2.75

INTRODUCTION TO NONLINEAR DIFFERENTIAL AND INTEGRAL EQUATIONS,
Harold T. Davis

Aspects of the problem of nonlinear equations, transformations that lead to equations solvable by classical means, results in special cases, and useful generalizations. Thorough, but easily followed by mathematically sophisticated reader who knows little about non-linear equations. 137 problems for student to solve. xv + 566pp. 5⅜ x 8½. 60971-5 Paperbound $2.75

AN INTRODUCTION TO THE GEOMETRY OF N DIMENSIONS, *D. H. Y. Sommerville*

An introduction presupposing no prior knowledge of the field, the only book in English devoted exclusively to higher dimensional geometry. Discusses fundamental ideas of incidence, parallelism, perpendicularity, angles between linear space; enumerative geometry; analytical geometry from projective and metric points of view; polytopes; elementary ideas in analysis situs; content of hyper-spacial figures. Bibliography. Index. 60 diagrams. 196pp. 5⅜ x 8.
60494-2 Paperbound $1.50

ELEMENTARY CONCEPTS OF TOPOLOGY, *P. Alexandroff*

First English translation of the famous brief introduction to topology for the beginner or for the mathematician not undertaking extensive study. This unusually useful intuitive approach deals primarily with the concepts of complex, cycle, and homology, and is wholly consistent with current investigations. Ranges from basic concepts of set-theoretic topology to the concept of Betti groups. "Glowing example of harmony between intuition and thought," David Hilbert. Translated by A. E. Farley. Introduction by D. Hilbert. Index. 25 figures. 73pp. 5⅜ x 8. 60747-X Paperbound $1.25

ELEMENTS OF NON-EUCLIDEAN GEOMETRY, *D. M. Y. Sommerville*

Unique in proceeding step-by-step, in the manner of traditional geometry. Enables the student with only a good knowledge of high school algebra and geometry to grasp elementary hyperbolic, elliptic, analytic non-Euclidean geometries; space curvature and its philosophical implications; theory of radical axes; homothetic centres and systems of circles; parataxy and parallelism; absolute measure; Gauss' proof of the defect area theorem; geodesic representation; much more, all with exceptional clarity. 126 problems at chapter endings provide progressive practice and familiarity. 133 figures. Index. xvi + 274pp. 5⅜ x 8. 60460-8 Paperbound $2.00

INTRODUCTION TO THE THEORY OF NUMBERS, *L. E. Dickson*

Thorough, comprehensive approach with adequate coverage of classical literature, an introductory volume beginners can follow. Chapters on divisibility, congruences, quadratic residues & reciprocity. Diophantine equations, etc. Full treatment of binary quadratic forms without usual restriction to integral coefficients. Covers infinitude of primes, least residues. Fermat's theorem. Euler's phi function, Legendre's symbol, Gauss's lemma, automorphs, reduced forms, recent theorems of Thue & Siegel, many more. Much material not readily available elsewhere. 239 problems. Index. I figure. viii + 183pp. 5⅜ x 8.
60342-3 Paperbound $1.75

MATHEMATICAL TABLES AND FORMULAS, *compiled by Robert D. Carmichael and Edwin R. Smith*

Valuable collection for students, etc. Contains all tables necessary in college algebra and trigonometry, such as five-place common logarithms, logarithmic sines and tangents of small angles, logarithmic trigonometric functions, natural trigonometric functions, four-place antilogarithms, tables for changing from sexagesimal to circular and from circular to sexagesimal measure of angles, etc. Also many tables and formulas not ordinarily accessible, including powers, roots, and reciprocals, exponential and hyperbolic functions, ten-place logarithms of prime numbers, and formulas and theorems from analytical and elementary geometry and from calculus. Explanatory introduction. viii + 269pp. 5⅜ x 8½. 60111-0 Paperbound $1.50

A SOURCE BOOK IN MATHEMATICS,
D. E. Smith
Great discoveries in math, from Renaissance to end of 19th century, in English translation. Read announcements by Dedekind, Gauss, Delamain, Pascal, Fermat, Newton, Abel, Lobachevsky, Bolyai, Riemann, De Moivre, Legendre, Laplace, others of discoveries about imaginary numbers, number congruence, slide rule, equations, symbolism, cubic algebraic equations, non-Euclidean forms of geometry, calculus, function theory, quaternions, etc. Succinct selections from 125 different treatises, articles, most unavailable elsewhere in English. Each article preceded by biographical introduction. Vol. I: Fields of Number, Algebra. Index. 32 illus. 338pp. 5⅜ x 8. Vol. II: Fields of Geometry, Probability, Calculus, Functions, Quaternions. 83 illus. 432pp. 5⅜ x 8.
60552-3, 60553-1 Two volume set, paperbound $5.00

FOUNDATIONS OF PHYSICS,
R. B. Lindsay & H. Margenau
Excellent bridge between semi-popular works & technical treatises. A discussion of methods of physical description, construction of theory; valuable for physicist with elementary calculus who is interested in ideas that give meaning to data, tools of modern physics. Contents include symbolism; mathematical equations; space & time foundations of mechanics; probability; physics & continua; electron theory; special & general relativity; quantum mechanics; causality. "Thorough and yet not overdetailed. Unreservedly recommended," *Nature* (London). Unabridged, corrected edition. List of recommended readings. 35 illustrations. xi + 537pp. 5⅜ x 8. 60377-6 Paperbound $3.50

FUNDAMENTAL FORMULAS OF PHYSICS,
ed. by D. H. Menzel
High useful, full, inexpensive reference and study text, ranging from simple to highly sophisticated operations. Mathematics integrated into text—each chapter stands as short textbook of field represented. Vol. 1: Statistics, Physical Constants, Special Theory of Relativity, Hydrodynamics, Aerodynamics, Boundary Value Problems in Math, Physics, Viscosity, Electromagnetic Theory, etc. Vol. 2: Sound, Acoustics, Geometrical Optics, Electron Optics, High-Energy Phenomena, Magnetism, Biophysics, much more. Index. Total of 800pp. 5⅜ x 8.
60595-7, 60596-5 Two volume set, paperbound $4.75

THEORETICAL PHYSICS,
A. S. Kompaneyets
One of the very few thorough studies of the subject in this price range. Provides advanced students with a comprehensive theoretical background. Especially strong on recent experimentation and developments in quantum theory. Contents: Mechanics (Generalized Coordinates, Lagrange's Equation, Collision of Particles, etc.), Electrodynamics (Vector Analysis, Maxwell's equations, Transmission of Signals, Theory of Relativity, etc.), Quantum Mechanics (the Inadequacy of Classical Mechanics, the Wave Equation, Motion in a Central Field, Quantum Theory of Radiation, Quantum Theories of Dispersion and Scattering, etc.), and Statistical Physics (Equilibrium Distribution of Molecules in an Ideal Gas, Boltzmann Statistics, Bose and Fermi Distribution. Thermodynamic Quantities, etc.). Revised to 1961. Translated by George Yankovsky, authorized by Kompaneyets. 137 exercises. 56 figures. 529pp. 5⅜ x 8½.
60972-3 Paperbound $3.50

MATHEMATICAL PHYSICS, *D. H. Menzel*
Thorough one-volume treatment of the mathematical techniques vital for classical mechanics, electromagnetic theory, quantum theory, and relativity. Written by the Harvard Professor of Astrophysics for junior, senior, and graduate courses, it gives clear explanations of all those aspects of function theory, vectors, matrices, dyadics, tensors, partial differential equations, etc., necessary for the understanding of the various physical theories. Electron theory, relativity, and other topics seldom presented appear here in considerable detail. Scores of definition, conversion factors, dimensional constants, etc. "More detailed than normal for an advanced text . . . excellent set of sections on Dyadics, Matrices, and Tensors," *Journal of the Franklin Institute*. Index. 193 problems, with answers. x + 412pp. 5⅜ x 8. 60056-4 Paperbound $2.50

THE THEORY OF SOUND, *Lord Rayleigh*
Most vibrating systems likely to be encountered in practice can be tackled successfully by the methods set forth by the great Nobel laureate, Lord Rayleigh. Complete coverage of experimental, mathematical aspects of sound theory. Partial contents: Harmonic motions, vibrating systems in general, lateral vibrations of bars, curved plates or shells, applications of Laplace's functions to acoustical problems, fluid friction, plane vortex-sheet, vibrations of solid bodies, etc. This is the first inexpensive edition of this great reference and study work. Bibliography, Historical introduction by R. B. Lindsay. Total of 1040pp. 97 figures. 5⅜ x 8. 60292-3, 60293-1 Two volume set, paperbound $6.00

HYDRODYNAMICS, *Horace Lamb*
Internationally famous complete coverage of standard reference work on dynamics of liquids & gases. Fundamental theorems, equations, methods, solutions, background, for classical hydrodynamics. Chapters include Equations of Motion, Integration of Equations in Special Gases, Irrotational Motion, Motion of Liquid in 2 Dimensions, Motion of Solids through Liquid-Dynamical Theory, Vortex Motion, Tidal Waves, Surface Waves, Waves of Expansion, Viscosity, Rotating Masses of Liquids. Excellently planned, arranged; clear, lucid presentation. 6th enlarged, revised edition. Index. Over 900 footnotes, mostly bibliographical. 119 figures. xv + 738pp. 6⅛ x 9¼. 60256-7 Paperbound $4.00

DYNAMICAL THEORY OF GASES, *James Jeans*
Divided into mathematical and physical chapters for the convenience of those not expert in mathematics, this volume discusses the mathematical theory of gas in a steady state, thermodynamics, Boltzmann and Maxwell, kinetic theory, quantum theory, exponentials, etc. 4th enlarged edition, with new material on quantum theory, quantum dynamics, etc. Indexes. 28 figures. 444pp. 6⅛ x 9¼. 60136-6 Paperbound $2.75

THERMODYNAMICS, *Enrico Fermi*
Unabridged reproduction of 1937 edition. Elementary in treatment; remarkable for clarity, organization. Requires no knowledge of advanced math beyond calculus, only familiarity with fundamentals of thermometry, calorimetry. Partial Contents: Thermodynamic systems; First & Second laws of thermodynamics; Entropy; Thermodynamic potentials: phase rule, reversible electric cell; Gaseous reactions: van't Hoff reaction box, principle of LeChatelier; Thermodynamics of dilute solutions: osmotic & vapor pressures, boiling & freezing points; Entropy constant. Index. 25 problems. 24 illustrations. x + 160pp. 5⅜ x 8. 60361-X Paperbound $2.00

CELESTIAL OBJECTS FOR COMMON TELESCOPES,
Rev. T. W. Webb
Classic handbook for the use and pleasure of the amateur astronomer. Of inestimable aid in locating and identifying thousands of celestial objects. Vol I, The Solar System: discussions of the principle and operation of the telescope, procedures of observations and telescope-photography, spectroscopy, etc., precise location information of sun, moon, planets, meteors. Vol. II, The Stars: alphabetical listing of constellations, information on double stars, clusters, stars with unusual spectra, variables, and nebulae, etc. Nearly 4,000 objects noted. Edited and extensively revised by Margaret W. Mayall, director of the American Assn. of Variable Star Observers. New Index by Mrs. Mayall giving the location of all objects mentioned in the text for Epoch 2000. New Precession Table added. New appendices on the planetary satellites, constellation names and abbreviations, and solar system data. Total of 46 illustrations. Total of xxxix + 606pp. 5⅜ x 8. 20917-2, 20918-0 Two volume set, paperbound $5.00

PLANETARY THEORY,
E. W. Brown and C. A. Shook
Provides a clear presentation of basic methods for calculating planetary orbits for today's astronomer. Begins with a careful exposition of specialized mathematical topics essential for handling perturbation theory and then goes on to indicate how most of the previous methods reduce ultimately to two general calculation methods: obtaining expressions either for the coordinates of planetary positions or for the elements which determine the perturbed paths. An example of each is given and worked in detail. Corrected edition. Preface. Appendix. Index. xii + 302pp. 5⅜ x 8½. 61133-7 Paperbound $2.25

STAR NAMES AND THEIR MEANINGS,
Richard Hinckley Allen
An unusual book documenting the various attributions of names to the individual stars over the centuries. Here is a treasure-house of information on a topic not normally delved into even by professional astronomers; provides a fascinating background to the stars in folk-lore, literary references, ancient writings, star catalogs and maps over the centuries. Constellation-by-constellation analysis covers hundreds of stars and other asterisms, including the Pleiades, Hyades, Andromedan Nebula, etc. Introduction. Indices. List of authors and authorities. xx + 563pp. 5⅜ x 8½. 21079-0 Paperbound $3.00

A SHORT HISTORY OF ASTRONOMY, *A. Berry*
Popular standard work for over 50 years, this thorough and accurate volume covers the science from primitive times to the end of the 19th century. After the Greeks and the Middle Ages, individual chapters analyze Copernicus, Brahe, Galileo, Kepler, and Newton, and the mixed reception of their discoveries. Post-Newtonian achievements are then discussed in unusual detail: Halley, Bradley, Lagrange, Laplace, Herschel, Bessel, etc. 2 Indexes. 104 illustrations, 9 portraits. xxxi + 440pp. 5⅜ x 8. 20210-0 Paperbound $2.75

SOME THEORY OF SAMPLING, *W. E. Deming*
The purpose of this book is to make sampling techniques understandable to and useable by social scientists, industrial managers, and natural scientists who are finding statistics increasingly part of their work. Over 200 exercises, plus dozens of actual applications. 61 tables. 90 figs. xix + 602pp. 5⅜ x 8½. 61755-6 Paperbound $3.50

PRINCIPLES OF STRATIGRAPHY,
A. W. Grabau
Classic of 20th century geology, unmatched in scope and comprehensiveness. Nearly 600 pages cover the structure and origins of every kind of sedimentary, hydrogenic, oceanic, pyroclastic, atmoclastic, hydroclastic, marine hydroclastic, and bioclastic rock; metamorphism; erosion; etc. Includes also the constitution of the atmosphere; morphology of oceans, rivers, glaciers; volcanic activities; faults and earthquakes; and fundamental principles of paleontology (nearly 200 pages). New introduction by Prof. M. Kay, Columbia U. 1277 bibliographical entries. 264 diagrams. Tables, maps, etc. Two volume set. Total of xxxii + 1185pp. 5⅜ x 8. 60686-4, 60687-2 Two volume set, paperbound $6.25

SNOW CRYSTALS, *W. A. Bentley and W. J. Humphreys*
Over 200 pages of Bentley's famous microphotographs of snow flakes—the product of painstaking, methodical work at his Jericho, Vermont studio. The pictures, which also include plates of frost, glaze and dew on vegetation, spider webs, windowpanes; sleet; graupel or soft hail, were chosen both for their scientific interest and their aesthetic qualities. The wonder of nature's diversity is exhibited in the intricate, beautiful patterns of the snow flakes. Introductory text by W. J. Humphreys. Selected bibliography. 2,453 illustrations. 224pp. 8 x 10¼. 20287-9 Paperbound $3.25

THE BIRTH AND DEVELOPMENT OF THE GEOLOGICAL SCIENCES,
F. D. Adams
Most thorough history of the earth sciences ever written. Geological thought from earliest times to the end of the 19th century, covering over 300 early thinkers & systems: fossils & their explanation, vulcanists vs. neptunists, figured stones & paleontology, generation of stones, dozens of similar topics. 91 illustrations, including medieval, renaissance woodcuts, etc. Index. 632 footnotes, mostly bibliographical. 511pp. 5⅜ x 8. 20005-1 Paperbound $2.75

ORGANIC CHEMISTRY, *F. C. Whitmore*
The entire subject of organic chemistry for the practicing chemist and the advanced student. Storehouse of facts, theories, processes found elsewhere only in specialized journals. Covers aliphatic compounds (500 pages on the properties and synthetic preparation of hydrocarbons, halides, proteins, ketones, etc.), alicyclic compounds, aromatic compounds, heterocyclic compounds, organophosphorus and organometallic compounds. Methods of synthetic preparation analyzed critically throughout. Includes much of biochemical interest. "The scope of this volume is astonishing," *Industrial and Engineering Chemistry*. 12,000-reference index. 2387-item bibliography. Total of x + 1005pp. 5⅜ x 8. 60700-3, 60701-1 Two volume set, paperbound $4.50

THE PHASE RULE AND ITS APPLICATION,
Alexander Findlay
Covering chemical phenomena of 1, 2, 3, 4, and multiple component systems, this "standard work on the subject" (*Nature,* London), has been completely revised and brought up to date by A. N. Campbell and N. O. Smith. Brand new material has been added on such matters as binary, tertiary liquid equilibria, solid solutions in ternary systems, quinary systems of salts and water. Completely revised to triangular coordinates in ternary systems, clarified graphic representation, solid models, etc. 9th revised edition. Author, subject indexes. 236 figures. 505 footnotes, mostly bibliographic. xii + 494pp. 5⅜ x 8.
60091-2 Paperbound $2.75

A COURSE IN MATHEMATICAL ANALYSIS,
Edouard Goursat

Trans. by E. R. Hedrick, O. Dunkel, H. G. Bergmann. Classic study of fundamental material thoroughly treated. Extremely lucid exposition of wide range of subject matter for student with one year of calculus. Vol. 1: Derivatives and differentials, definite integrals, expansions in series, applications to geometry. 52 figures, 556pp. 60554-X Paperbound $3.00. Vol. 2, Part I: Functions of a complex variable, conformal representations, doubly periodic functions, natural boundaries, etc. 38 figures, 269pp. 60555-8 Paperbound $2.25. Vol. 2, Part II: Differential equations, Cauchy-Lipschitz method, nonlinear differential equations, simultaneous equations, etc. 308pp. 60556-6 Paperbound $2.50. Vol. 3, Part I: Variation of solutions, partial differential equations of the second order. 15 figures, 339pp. 61176-0 Paperbound $3.00. Vol. 3, Part II: Integral equations, calculus of variations. 13 figures, 389pp. 61177-9 Paperbound $3.00 60554-X, 60555-8, 60556-6 61176-0, 61177-9 Six volume set, paperbound $13.75

PLANETS, STARS AND GALAXIES,
A. E. Fanning

Descriptive astronomy for beginners: the solar system; neighboring galaxies; seasons; quasars; fly-by results from Mars, Venus, Moon; radio astronomy; etc. all simply explained. Revised up to 1966 by author and Prof. D. H. Menzel, former Director, Harvard College Observatory. 29 photos, 16 figures. 189pp. 5⅜ x 8½. 21680-2 Paperbound $1.50

GREAT IDEAS IN INFORMATION THEORY, LANGUAGE AND CYBERNETICS,
Jagjit Singh

Winner of Unesco's Kalinga Prize covers language, metalanguages, analog and digital computers, neural systems, work of McCulloch, Pitts, von Neumann, Turing, other important topics. No advanced mathematics needed, yet a full discussion without compromise or distortion. 118 figures. ix + 338pp. 5⅜ x 8½. 21694-2 Paperbound $2.25

GEOMETRIC EXERCISES IN PAPER FOLDING,
T. Sundara Row

Regular polygons, circles and other curves can be folded or pricked on paper, then used to demonstrate geometric propositions, work out proofs, set up well-known problems. 89 illustrations, photographs of actually folded sheets. xii + 148pp. 5⅜ x 8½. 21594-6 Paperbound $1.00

VISUAL ILLUSIONS, THEIR CAUSES, CHARACTERISTICS AND APPLICATIONS,
M. Luckiesh

The visual process, the structure of the eye, geometric, perspective illusions, influence of angles, illusions of depth and distance, color illusions, lighting effects, illusions in nature, special uses in painting, decoration, architecture, magic, camouflage. New introduction by W. H. Ittleson covers modern developments in this area. 100 illustrations. xxi + 252pp. 5⅜ x 8. 21530-X Paperbound $1.50

ATOMS AND MOLECULES SIMPLY EXPLAINED,
B. C. Saunders and R. E. D. Clark

Introduction to chemical phenomena and their applications: cohesion, particles, crystals, tailoring big molecules, chemist as architect, with applications in radioactivity, color photography, synthetics, biochemistry, polymers, and many other important areas. Non technical. 95 figures. x + 299pp. 5⅜ x 8½. 21282-3 Paperbound $1.50